Lecture Notes in Mathematics

A collection of informal reports and seminars
Edited by A. Dold, Heidelberg and B. Eckmann, Zürich

Series: Tata Institute of Fundamental Research, Bombay
Adviser: M. S. Narasimhan

T0222661

175

Marcel Brelot

Université de Paris, Paris / France

On Topologies and Boundaries in Potential Theory

(Enlarged edition of a course of lectures delivered in 1966)

Springer-Verlag
Berlin · Heidelberg · New York 1971

ISBN 3-540-05327-1 Springer-Verlag Berlin · Heidelberg · New York
ISBN 0-387-05327-1 Springer-Verlag New York · Heidelberg · Berlin

Offsetdruck: Julius Beltz, Weinheim/Bergstr.

Introduction

Potential theory, whose rich structure has provoked much research in such fields as capacity, distributions, extreme elements, Dirichlet spaces, Hunt's kernels and semi-groups, probability theory etc., has also led to the introduction of new topologies and boundaries. Among these notions' chose which enable us to express general results on the behaviour of potential functions will be especially studied here. Their success is due to the fact that they are adapted to the nature of the functions studied. Classical potential theory introduced a notion of thinness in Euclidean n-space $\mathbb{R}^n$ (Brelot [4,5]) and a corresponding "fine" topology (H. Cartan [2]) for which the potential functions are continuous. It also introduced several boundaries, chiefly the Martin boundary (see the book of Constantinescu-Cornea [1]) and a notion of thinness at this boundary (Naim [1]), which led Doob [3,4] to definitive results generalizing the famous Fatou theorems on radial or non-tangential limits. It was possible to extend these notions to axiomatic potential theories, which include the theory of second-order partial differential equations of elliptic type and to some extent also those of parabolic type. Probabilistic interpretations and connections with Markov processes were also developed.

The importance of such topological researches in potential theory justifies their further study. This course of lectures will be an introduction to this vast subject. The rich book of Constantinescu-Cornea constituted an important synthesis of classical potential theory on Riemann surfaces, particularly regarding boundary questions. Here, I shall start from general, abstract, axiomatic ideas and then give some important applications to pure and potential theory and function theory, mainly in the classical case (on $\mathcal{E}$-spaces or Green spaces which include Riemann surfaces). Arguments will generally be chosen with a view to possible extension or adaptation to axiomatic theories of harmonic functions, but these will be mentioned only briefly. Probabilistic interpretations will be omitted (for this field, see Doob [1, 2,3], Hunt [1] and the books of Dynkin [1], P.A.Meyer [1,2], Blumenthal-Getoor [1]).

The first part of the lectures will deal with the notions of (inner or internal) thinness and fine topology, corresponding to a given space and a cone of positive functions. These will be illustrated in applications, such as refinements of classical balayage theory.

The second part will study boundaries: first, an abstract "minimal boundary" (consisting of 'extreme' elements in the set of functions) with a concept of minimal thinness, and then, classical or axiomatic potential theory, the famous Martin boundary and its "minimal" part, where the minimal thinness is the key to the fundamental boundary behaviour. It is curious that a topology corresponding to this thinness may be interpreted in most applications as the inner fine topology (dealt with in Part I) on the space completed by the minimal boundary, for a suitable new set of functions on the whole space.

A detailed bibliography is included in order to facilitate further study. (Authors' names are listed in alphabetical order and each author's papers are numbered). Definitions, lemmas, propositions and theorems will be numbered in the same serial order in each chapter. For brevity of notation, we shall often denote by the same symbol a point and the set defined by this point.

I am grateful to Mr. S. Ramaswamy for his valuable help in the preparation of this volume.

Table of Contents

FIRST PART

Inner Fine Topology

SECOND PART

Boundary Theories and Minimal Thinness

CHAPTER I

General Notions of Thinness and Fine Topology[1]

1. Let Ω be any set, not necessarily a topological space. Let Φ be any set of extended real-valued non-negative functions on Ω forming a convex cone i.e. satisfying the following conditions:

i) $f \in \Phi \Rightarrow \lambda f \in \Phi$, $\forall \lambda$ real and ≥ 0

ii) $f_1, f_2 \in \Phi \Rightarrow f_1 + f_2 \in \Phi$ and moreover,

iii) $+\infty \in \Phi$.

We make it a convention that $0 . \infty = 0$ so that i) is meaningful when $\lambda = 0$ and f takes the value $+\infty$ at some point $x \in \Omega$. We shall suppose sometimes that Φ is C-closed i.e. that whenever $f_n \in \Phi$, $n = 1,2,\dots$ Σf_n also belongs to Φ (countable addivity condition). But this will be always explicitly mentioned.

If Ω has no topology, we may introduce the coarsest topology $\mathcal{C}_0$ making all the functions $u \in \Phi$ lower semi-continuous. This means that the topology $\mathcal{C}_0$ is generated by sets of the form $\{ x \mid u(x) > a \}$ where u is any function in Φ and a is any real number. In case Ω is already a topological space with topology $\mathcal{C}_1$ to start with, as it is in most applications, we impose one more condition on Φ namely that the functions belonging to Φ are lower semi-continuous.

__Remark__. In a topological space, let us say that an extended real-valued function f has a __global peak point__ at x_0, if for any neighbourhood $\mathcal{V}$ of x_0, $\sup_{C\mathcal{V}} f < f(x_0)$.

If we suppose with the topology $\mathcal{C}_1$, that any $x_0 \in \Omega$ is a global peak point for some $v \in \Phi$, then $\mathcal{C}_1 \equiv \mathcal{C}_0$.

Because, for any $\mathcal{C}_1$- open set ω and any $x_0 \in \omega$, for a suitable

---

(1) This axiomatic introduction is adapted from the paper (Brelot [21]).

$v \in \Phi$, $\sup_{\complement\omega} v < \lambda < v(x_0)$ and the set $\{v > \lambda\}$ is an open $\mathcal{T}_0$-neighbourhood of x_0 contained in ω.

Fine topology. In Ω, we introduce another topology $\mathcal{T}$ which is finer than $\mathcal{T}_1$ and the coarsest for which all the functions $u \in \Phi$ are continuous. Since sets of the form $\{x \mid u(x) > \lambda\}$ are already open in $\mathcal{T}_1$, $\mathcal{T}$ is the topology whose class of open sets is generated by

i) all the open sets in $\mathcal{T}_1$

ii) sets of the form $\{x \mid u(x) < \lambda\}$ where $u \in \Phi$ and λ any real number.

Definition I, 1. The topology $\mathcal{T}$ is called the fine topology in Ω (and is said to be associated to Φ and $\mathcal{T}_1$ in Ω)[short form for internal or inner fine topology .

Definition I, 2. A subset e of Ω is said to be thin at $x_0 \notin e$ (2) (i.e. inner or internal thin)

if: i) $x_0 \notin \bar{e}$ in $\mathcal{T}_1$

or ii) if $x_0 \in \bar{e}$, $\exists u \in \Phi$ s.t

$u(x_0) < \liminf_{\substack{x \in e \\ x \to x_0}} u(x)$ where the right member is $\sup_{\sigma} (\inf_{x \in e \cap \sigma} u(x))$, σ being a neighbourhood of x_0.

According to the convention that the infimum of u over the void set is $+\infty$, the previous lim inf has always a meaning and the condition ii) for any x_0 contains i) by taking $u = 0$. Another definition is the existence of a $u \in \Phi$ and a neighbourhood σ of x_0 such that

$$u(x_0) < \inf_{x \in e \cap \sigma} u(x)$$

Note that e is not thin at $x_0 \notin e$ iff for every $u \in \Phi$

$u(x_0) = \liminf_{\substack{x \in e \\ x \to x_0}} u(x)$. We say that e is unthin at

(2) The case where $x_0 \in e$ will be introduced in Chap. V.

$x_0 \notin e$ if e is not thin at x_0.

2. <u>First Properties</u>:

i) Any subset of a thin set at x_0 is also thin at x_0.

ii) A finite union of thin sets is also thin at x_0.

iii) For a suitable neighbourhood σ of x_0, $e \cap \sigma$ is contained in an open set δ which is thin at $x_0 \notin \delta$.

We use the terms fine neighbourhood, fine closure (denoted by $\tilde{e}$ for the set e), fine limits etc. whenever we talk of neighbourhoods, closure, limit etc. in the fine topology.

<u>Theorem I, 3</u> (H. Cartan)(3)

<u>Fine neighbourhoods of</u> x_0 <u>are precisely the sets of the form</u> $\complement e$ <u>where</u> e <u>is a thin set at</u> x_0, $x_0 \notin e$.

<u>Proof</u>. Assume that e is thin at x_0. We shall prove that $\complement e$ contains a fine neighbourhood of x_0.

There exists a $u \in \Phi$ s.t $u(x_0) < \sup\limits_{\sigma} (\inf\limits_{x \in e \cap \sigma} u)$. Therefore, there exists a suitable neighbourhood σ of x_0 such that $u(x_0) < \inf\limits_{x \in e \cap \sigma} u(x)$

Consider the set $E = \{x \mid u(x) < \lambda\}$. Then $E \ni x_0$ and E is open in $\mathcal{C}$. Also since $u > \lambda$ on $e \cap \sigma$, $E \cap e \cap \sigma = \emptyset$ i.e. $E \cap \sigma \subset \complement e$. Now, σ and E are neighbourhoods of x_0 in $\mathcal{C}$. Therefore, $\complement e$ is a fine neighbourhood of x_0.

Conversely, assume that $\mathcal{V}$ is a fine neighbourhood of x_0. We shall prove that $\complement \mathcal{V}$ is thin at x_0.

$\mathcal{V}$ contains a fine open set E containing x_0. Suppose E is $\mathcal{C}_1$-open (i.e. open in the topology $\mathcal{C}_1$), then $x_0 \notin \overline{\complement \mathcal{V}}$. Therefore, $\complement \mathcal{V}$ is thin at x_0 by i). If E is not $\mathcal{C}_1$-open, then E is the intersection of a $\mathcal{C}_1$-open set (this set may be even Ω) with a finite number of sets of the form

---

(3) Due to H. Cartan in classical Potential Theory (See [2] Chap. VI).

$\{ x : u(x) < \lambda \}$ containing x_o. Their complementary sets are obviously thin at x_o. Since, finite union of thin sets is thin, E is also thin at x_o. Since $\complement \mathcal{V} \subset \complement E$, $\complement \mathcal{V}$ is also thin at x_o.

Remarks. i) A fine isolated point of e is a point x of e such that $e - \{x\}$ is thin at x.

ii) $\tilde{e}$ (fine closure of e) = e $\cup$ $\{$ points of $\complement e$ where e is not thin $\}$.

Exercises.

1) A base of filtre $\mathcal{B}$ converging to x_o is said to be thin at x_o if there exists a $u \in \Phi$ such that $u(x_o) < \liminf_{\mathcal{B}} u = \sup_{\sigma \in \mathcal{B}} (\inf_{\sigma} u)$. Show that

α) The thinness of e at $x_o \notin e$, $x_o \in \bar{e}$ means the thinness of the base of the filtre defined by the intersections of e with the neighbourhoods of x_o.

β) A criterion of thinness of $\mathcal{B}$ is the existence of $e \in \mathcal{B}$, $e \not\ni x_o$, e thin at x_o.

Exercises

2) (H. Bauer) Consider a convex set Ω of a Hausdorff topological vector space and the set Φ of the finite lower semi-continuous concave functions ≥ 0 on Ω completed by the function $+\infty$. The following properties are equivalent.

α) x_o is an extreme point of Ω.

β) $\complement\{x_o\}$ is thin at x_o (even strongly thin, see Chap. II).

γ) The inequality $u_1 \leq u_2$ of two finite functions of Φ on $\Omega \setminus \{x_o\}$ does not imply the same inequality at x_o.

3. Some properties of the fine topology.

Theorem I, 4. a) If $\mathcal{T}_1$ is Hausdorff, so is $\mathcal{T}$.

b) If $\mathcal{T}_1$ is regular, so is $\mathcal{T}$ (in fact, even if $\mathcal{T}_1$ is not Hausdorff, if any $\mathcal{T}_1$ neighbourhood of x_o contains a closed neighbourhood, any fine neighbourhood of x_o contains a fine neighbourhood which is even $\mathcal{C}_1$-closed).

<u>If $\mathcal{C}_1$ is locally compact</u>(4), <u>the $\mathcal{C}$-space is a Baire space</u>.

c) <u>If $\mathcal{C}_1$ is uniformisable so is $\mathcal{C}$</u>.

<u>Proof</u>. a) is obvious because $\mathcal{C}$ is finer than $\mathcal{C}_1$.

b) We consider a fine neighbourhood $\mathcal{V}$ of x_o; there exists a $u \in \Phi$ such that for a suitable $\mathcal{C}_1$-neighbourhood σ of x_o, $u(x_o) < \inf_{y \in \complement \mathcal{V} \cap \sigma} u(y)$.

Let λ be any real number strictly between both numbers and $E = \left\{ x \,\middle|\, u(x) \leq \lambda \right\}$. E is $\mathcal{C}_1$-closed and is a fine neighbourhood of x_o; moreover $E \cap \sigma \subset \mathcal{V}$. Now, according to the hypothesis, there exists a $\mathcal{C}_1$-neighbourhood σ_1 of x_o which is closed and contained in σ; $E \cap \sigma_1$ is $\mathcal{C}_1$-closed, is a fine neighbourhood of x_o and is contained in $E \cap \sigma \subset \mathcal{V}$.

Suppose now $\mathcal{C}_1$ is locally compact. Consider a sequence of fine open sets ω_n which are fine dense in Ω and let us prove that $\bigcap \omega_n$ is fine dense i.e. that for any fine open $\omega \neq \emptyset$, $\bigcap \omega_n \cap \omega \neq \emptyset$.

In $\omega_1 \cap \omega$ which is not empty, we choose a point x_1 and a fine neighbourhood of x_1 which is a $\mathcal{C}_1$-closed set, then even a $\mathcal{C}_1$-compact set K_1 (by intersection with a compact neighbourhood). The fine interior α_1 of K_1 is not empty.

In $\omega_2 \cap \alpha_1$ we choose in the same way a compact set K_2 whose fine interior is $\alpha_2 \neq \emptyset$ .... We get $\omega \supset K_1 \supset \alpha_1 \supset \omega_2 \cap \alpha_1 \supset K_2 \supset \alpha_2 \supset \omega_3 \cap \alpha_2 \ldots$. Hence $\omega_n \cap \omega \supset \omega_n \cap \alpha_{n-1} \supset K_n \supset \bigcap_i K_i$ and $(\bigcap_n \omega_n) \cap \omega = \bigcap_n (\omega_n \cap \omega) \supset \bigcap_i K_i \neq \emptyset$.

c) For this, it is sufficient to prove the following:

Suppose $\mathcal{V}$ is any fine open set containing x_o. Then there exists a fine continuous function equal to zero at x_o and to 1 in $\complement \mathcal{V}$ such that the

---

(4) or only if any $\mathcal{C}_1$-neighbourhood of any x_o contains a $\mathcal{C}_1$-neighbourhood which is semi-compact (i.e. satisfying the covering property, but perhaps not Hausdorff). Similar Proof.

values of this function lie in $[0, 1]$.

As $\complement\mathcal{V}$ is thin at x_0, there exists a $u \in \Phi$ and a suitable neighbourhood σ of x_0 such that

$$\inf_{\complement\mathcal{V} \cap \sigma} u = \lambda > u(x_0).$$

Let $u_1 = \inf(u, \lambda)$. Then u_1 is fine continuous.

Let $u_2 = \frac{1}{\lambda - u(x_0)} \left[\sup(u_1, u(x_0)) - u(x_0) \right]$.

Then u_2 is fine continuous with values in $[0, 1]$. Moreover,

$$u_2 = 0 \text{ at } x_0$$
$$= 1 \text{ in } \complement\mathcal{V} \cap \sigma.$$

Since $\mathcal{T}_1$ is uniformisable there exists a ω, continuous from Ω to $[0, 1]$ such that $\omega(x_0) = 0$ and $\omega(x) = 1$ on $\complement\sigma$.

Let $f = \sup(u_2, \omega)$. Then f is fine continuous

$= 0$ at x_0

$= 1$ on $\complement\mathcal{V}$ and $0 \leq f(x) \leq 1$.

4. Definition I, 5.

a) A set e is said to be hyperthin at $x_0 \notin e$, if there exists a $u \in \Phi$ such that $u(x_0)$ is finite, $u(x) \to +\infty$, $x \in e$ $x \to x_0$.

b) e is said to be polar if there exists a $u \in \Phi$ such that $u = +\infty$ on e, and $u \not\equiv +\infty$.

c) e is said to be strictly polar if for any non-void subset e' of e and any $x \notin e'$, there exists a $u \in \Phi$ such that $u = +\infty$ on e' and $u(x) < +\infty$.

A point x is said to be polar, strictly polar if $\{x\}$ is so.

Proposition I, 6. If e is strictly polar, $e \setminus \{x\}$ is hyperthin at x, $\forall x$.

Some negative properties of the fine topology.

The fine topology is not easy to work with, like the common topologies.

In many applications, the following conditions are satisfied and show the difficulties.

As it will be commonly realized in the applications, suppose Φ is C-closed, we first remark that for any countable set e of strictly polar points, $e \sim \{x_o\}$ is hyperthin at x_o for every x_o.

If the points of e different from x_o are denoted x_n, and if $u_n \in \Phi$ are such that $u_n(x_n) = +\infty$, $u_n(x_o) < \frac{1}{2^n}$, then $v = \sum_1 u_n \in \Phi$, $v = +\infty$ on $e - \{x_o\}$, $v(x_o)$ is finite.

Let us emphasize:

<u>Proposition I, 7</u>. If Φ is C-closed, any sequence $\{x_n\}$ of strictly polar points has no fine limit point, unless there is a constant subsequence, and no fine limit unless x_n = constant for n large enough.

Suppose x_o is a cluster value; if there is no constant subsequence, $x_n \neq x_o$ for $n \geqslant N$ large enough; $\bigcup_N^{\infty} \{x_n\}$ is thin at x_o and the complementary set would be a fine neighbourhood of x_o without points x_n, $n \geqslant N$.

Suppose $x_n \to x_o$ and $x_n \neq x_o$ for an infinite set of n; there would be a subsequence $x_{n_p} \neq x_o$ and $\complement\{x_{n_p}\}$ would be a fine neighbourhood of x_o, which does not contain any x_n for n large enough.

<u>Exercises</u>. Suppose Φ is C-closed and that every point is strictly polar, then

i) Any countable set is fine closed

ii) No infinite set is fine quasi-compact

iii) If $\{x_o\}$ is not fine isolated, x_o has no countable base of fine neighbourhoods

iv) If the fine topology, even when it is Hausdorff, is not discrete, it is not metrizable.

5. <u>First general examples in Potential Theory</u>.

In $\mathbb{R}^n (n \geqslant 3)$, let us take for Φ, the set of all classical super harmonic non-negative functions and $+\infty$. $\mathcal{C}_1$ will be the Euclidean topology.

Using the particular superharmonic functions $h_{x_0}(x) = |x - x_0|^{2-n}$, it is obvious that any open set is a union of sets of type $h_{x_0} > \alpha$ and therefore $\mathcal{C}_0 = \mathcal{C}_1$. Every point is strictly polar and it is well known that if u_n is superharmonic, non-negative, $\sum u_n$ is superharmonic or $+\infty$ i.e. our Φ is closed under countable additions i.e. our Φ is C-closed.

It is also elementary that every polar set is strictly polar, therefore is hyperthin at every point outside and no point is fine isolated.

A non-trivial example of a thin set may be given for instance in $\mathbb{R}^3$ by the union α of balls $B_{x_n}^{r_n}$, $x_n \to x_0$ with r_n satisfying $\sum_{n=1}^{\infty} \dfrac{nr_n}{|x_n - x_0|}$, finite, $r_n < |x_n - x_0|$. The masses nr_n at the points x_n give a Newtonian potential (superharmonic > 0) finite at x_0, $\geq n$ on $B_{x_n}^{r_n}$ and therefore tending to $+\infty$ on α.

We will give at the beginning of Chapter IX, various elementary examples of thinness and unthinness that may be now looked at. But we will first continue to give general properties.

CHAPTER II

Notion of Reduced function. Applications.

Strong Thinness and Strong Unthinness

1. With the same space Ω and a cone Φ, consider a set $e \subset \Omega$ and let φ be an extended real-valued non-negative function on Ω. Define a function

$$R_\varphi^e \quad \text{as} \quad \inf_{\substack{u \in \Phi \\ u \geqslant \varphi \text{ on } e}} u \, .$$

This implies $R_\varphi^\emptyset = 0$.

Definition II, 1. This R_φ^e is called the reduced function(5) of φ on e.

When $e = \Omega$, we write R_φ instead of R_φ^Ω. It is easy to deduce the following properties of the reduced function.

i) $R_\varphi^e = R_{\varphi \cdot \chi_e}$ where χ_e is the characteristic function of e.

ii) $R_{\lambda\varphi}^e = \lambda R_\varphi^e \quad \forall \quad \lambda \geqslant 0.$

iii) $R_{\varphi_1 + \varphi_2}^e \leq R_{\varphi_1}^e + R_{\varphi_2}^e$ and

iv) $R_\varphi^{e_1 \cup e_2} \leqslant R_\varphi^{e_1} + R_\varphi^{e_2}$

v) If Φ is C-closed

$$R_\varphi^{\bigcup_{n=1}^{\infty} e_n} \leq \sum_{n=1}^{\infty} R_\varphi^{e_n} \, .$$

We shall give a characterisation of thin sets at a point x_0 involving reduced functions.

---

(5) This name was given by Brelot [17] for a similar notion introduced by Martin [1] for a boundary set. The definition of the text appeared in Brelot [14] and [20]. The function of the text after lower semi continuous regularisation (called balayaged function) was commonly used in classical Potential Theory already in Brelot [8] under the name of "extremale". See Chap. VI.(extremale relative to the complementary set)
For the developments of this chapter see Brelot [21].

Theorem II, 2. e is thin at x_0 iff $\operatorname{Inf}_{\sigma} R_1^{e\cap\sigma}(x_0) < 1$, σ denoting a variable neighbourhood of x_0, i.e. iff there exists a σ such that $R_1^{e\cap\sigma}(x_0) < 1$.

Proof. Suppose e is thin at x_0. Then there exists a neighbourhood σ of x_0 and a $u \in \Phi$ such that

$$u(x_0) < \inf_{x\in e\cap\sigma} u(x).$$

Let
$$u(x_0) < \lambda < \inf_{x\in e\cap\sigma} u(x).$$

This shows that $R_\lambda^{e\cap\sigma}(x) \leq u(x)$. Therefore

$$R_\lambda^{e\cap\sigma}(x_0) \leq u(x_0) < \lambda$$

$$\text{i.e. } R_1^{e\cap\sigma}(x_0) < 1.$$

Conversely, suppose there exists a neighbourhood σ of x_0 such that $R_1^{e\cap\sigma}(x_0) < 1$ then there exists a $u \in \Phi$ such that $u(x_0) < 1$ and $u \geq 1$ on $e\cap\sigma$. This implies that e is thin at x_0.

Proposition II, 3. If there exists a $u \in \Phi$ finite and > 0 at x_0, then for any set $e \not\ni x_0$, $\inf_{\sigma} R_1^{e\cap\sigma}(x_0) \leq 1$, σ denoting a variable neighbourhood of x_0.

Proof. It is sufficient to consider the case when e is not thin at x_0.

Therefore $u(x_0) = \sup_{\sigma} \inf_{e\cap\sigma} u(x)$. Therefore, given any λ such that $0 < \lambda < u(x_0)$, there exists a σ_0 such that

$$\lambda < \inf_{e\cap\sigma_0} u(x).$$

Hence,
$$R_\lambda^{e\cap\sigma_0} \leq u \quad\text{or}\quad \lambda R_1^{e\cap\sigma_0} \leq u.$$

Hence,
$$R_1^{e\cap\sigma_0}(x_0) \leq \frac{u(x_0)}{\lambda}.$$

Therefore $\operatorname{Inf}_{\sigma} R_1^{e\cap\sigma}(x_0) \leq \frac{u(x_0)}{\lambda}$. This being true for every $\lambda > 0$ such that $\lambda < u(x_0)$, we conclude that $\operatorname{Inf}_{\sigma} R_1^{e\cap\sigma}(x_0) \leq 1$.

Proposition II, 4. If φ is finite, continuous and > 0 at x_0, $x_0 \notin e$, then $\inf_{\sigma} R_\varphi^{e\cap\sigma} = \varphi(x_0) \inf_{\sigma} R_1^{e\cap\sigma}$. Therefore, a criterion for thinness is that

$\inf_{\sigma} R_{\varphi}^{e\cap\sigma}(x_0) < \varphi(x_0)$ for every φ finite, continuous and > 0 at x_0.

Proof. If $0 < \theta_1 < 1 < \theta_2$, then there exists a neighbourhood σ_0 of x_0 such that on $e \cap \sigma_0$

$$\theta_1 \varphi(x_0) \leqslant \varphi \leq \theta_2 \varphi(x_0) \text{ due to the continuity of } \varphi \text{ at } x_0.$$

Therefore, for any neighbourhood

$$\sigma \subset \sigma_0, \theta_1 \varphi(x_0)\, R_1^{e\cap\sigma} \leqslant R_{\varphi}^{e\cap\sigma} \leqslant \theta_2 \varphi(x_0)\, R_1^{e\cap\sigma}, \text{ then}$$

$$\theta_1 \varphi(x_0) \operatorname*{Inf}_{\sigma} R_1^{e\cap\sigma} \leqslant \operatorname*{Inf}_{\sigma} R_{\varphi}^{e\cap\sigma} \leqslant \theta_2 \varphi(x_0) \operatorname*{Inf}_{\sigma} R_1^{e\cap\sigma}$$

and because, θ_1, θ_2 are arbitrary,

$$\operatorname*{Inf}_{\sigma} R_{\varphi}^{e\cap\sigma} = \varphi(x_0) \operatorname*{Inf}_{\sigma} R_1^{e\cap\sigma} .$$

<u>Proposition II, 5</u>. If $\varphi \geqslant 0$ is lower semi continuous at x_0 and e is not thin at $x_0 \notin e$, then $R_{\varphi}^{e}(x_0) \geqslant \varphi(x_0)$ and if $\varphi \in \Phi$, $R_{\varphi}^{e}(x_0) = \varphi(x_0)$. For any φ lower-semi-continuous everywhere, not necessarily belonging to Φ, $R_{\varphi}^{e} = R_{\varphi}^{\tilde{e}}$.

Proof. If $v \in \Phi$ satisfies, $v \geqslant \varphi$ on e, then

$$v(x_0) = \liminf_{\substack{x\in e\\ x\to x_0}} v(x) \geqslant \liminf_{\substack{x\in e\\ x\to x_0}} \varphi(x) \geqslant \varphi(x_0).$$

Hence, $R_{\varphi}^{e}(x_0) \geqslant \varphi(x_0)$; if $\varphi \in \Phi$, we have also the opposite inequality and hence the equality.

Now, if φ is every where l.s.c. $v \geqslant \varphi$ on e, then $v \geqslant \varphi$ on $\tilde{e}$ as well. Hence $R_{\varphi}^{e} \geqslant R_{\varphi}^{\tilde{e}}$ and hence the equality.

<u>Proposition II, 6</u>. Consider a real-valued function $\varphi \geqslant 0$, satisfying $\varphi(x) \leqslant \varphi(x_0)$ finite, possessing a global peak point at x_0, and lower semi-continuous at x_0 (therefore, continuous at x_0). Suppose any constant > 0 is in Φ. Then the thinness of e at $x_0 \notin e$ is equivalent to $R_{\varphi}^{e}(x_0) < \varphi(x_0)$.

Proof. If e is not thin, we just saw the opposite inequality $\geqslant$. Suppose e is thin, there exists a $v \in \Phi$ and a neighbourhood σ of x_0 such that

$\inf_{\sigma \cap e} v > v(x_0)$. We choose λ strictly between these numbers and consider the function $\omega = \varphi(x_0) + k(v - \lambda)$, k a constant such that $0 < k < \frac{\varphi(x_0)}{\lambda}$.

$\omega \in \Phi$ and satisfies

$$\omega > \varphi(x_0) \geq \varphi \quad \text{on } e \cap \sigma$$

$\omega \geq \varphi(x_0) - k\lambda$ which majorizes φ on $\complement\sigma$ for k small enough. Therefore, $\omega \geq \varphi$ on e, $\omega \geq R^e_\varphi$ and

$$R^e_\varphi(x_0) \leq \omega(x_0) < \varphi(x_0).$$

Extension. Same conclusion if φ is the sum of a function φ_1 of the previous type and of a function $u \in \Phi$, finite at x_0.

Same argument for e not thin. If e is thin, $R^e_{\varphi_1}(x_0) < \varphi_1(x_0)$, $R^e_u(x_0) \leq u(x_0)$ and therefore

$$R^e_\varphi(x_0) \leq R^e_{\varphi_1}(x_0) + R^e_u(x_0) < \varphi_1(x_0) + u(x_0) = \varphi(x_0).$$

Remark. In the previous Propositions 5, 6, φ may be as well supposed to be fine l.s.c. instead of l.s.c.

3. Strong Thinness.

Definition II, 7. A set e is said to be strongly thin at a point $x_0 \notin e$ if

$$\operatorname*{Inf}_{\sigma} R_1^{e \cap \sigma}(x_0) = 0.$$

Properties. 1) A strongly thin set is thin.

2) Any subset of a strongly thin set is strongly thin.

3) A finite union of strongly thin sets is strongly thin.

Theorem II, 8. If given $\varepsilon > 0$, every function belonging to Φ finite at x_0 can be expressed as $v + \theta$ where θ is finite at x_0 and continuous, $v \in \Phi$ and $v(x_0) < \varepsilon$, then every set e thin at x_0 is also strongly thin at x_0.

Proof. Let e be thin at x_0. Then there exists a $u \in \Phi$ such that $u(x_0) < \sup_{\sigma} \inf_{x \in e \cap \sigma} u(x)$. Let K be any number such that $0 < K < \sup_{\sigma} \inf_{x \in e \cap \sigma} u(x) - u(x_0)$.

By hypothesis, there exists v and θ, $v \in \Phi$ with $v(x_0) < K\varepsilon$ (ε given > 0) and θ finite and continuous at x_0. Then

$$\sup_{\sigma} \inf_{x \in e \cap \sigma} u(x) - u(x_0)$$

$$= \sup_{\sigma} \inf_{x \in e \cap \sigma} v(x) - v(x_0).$$

As $0 < K < \sup_{\sigma} \inf_{x \in e \cap \sigma} v(x) - v(x_0)$,

$$\sup_{\sigma} \inf_{x \in e \cap \sigma} v(x) > K$$

and there exists a σ_0 such that

$$\inf_{x \in e \cap \sigma_0} v(x) > K.$$

Therefore, $R_K^{e \cap \sigma_0} \leq v$,

or $K\, R_1^{e \cap \sigma_0} \leq v$.

Hence $R_1^{e \cap \sigma_0}(x_0) \leq \frac{v(x_0)}{K} \leq \varepsilon$.

Therefore, $\inf_{\sigma} R_1^{e \cap \sigma}(x_0) \leq \varepsilon$.

Since this is true for every $\varepsilon > 0$, $\inf_{\sigma} R_1^{e \cap \sigma}(x_0) = 0$. Therefore, e is strongly thin at x_0.

Remark. The above theorem for e holds good even if we have such a decomposition in the form $v + \theta$, only for the function $u \in \Phi$ that is associated to e, i.e. for the one that satisfies

$$u(x_0) < \liminf_{\substack{x \to x_0 \\ x \in e}} u(x)$$

<u>Theorem II, 9</u>. <u>Assume that Φ is C-closed. Then, if e_n is a sequence of sets that are strongly thin at x_0, there exists a decreasing sequence σ_n of neighbourhoods of x_0 such that $\bigcup(e_n \cap \sigma_n)$ is strongly thin at x_0.</u>

Let ε_n be a sequence of positive real numbers such that $\sum \varepsilon_n < +\infty$. Since e_1 is strongly thin at x_0, there exists a neighbourhood σ_1 of x_0 such that $R_1^{e_1 \cap \sigma_1}(x_0) < \varepsilon_1$. Since e_2 is strongly thin at x_0, we can choose a neighbourhood σ_2 of x_0 $\sigma_2 \subset \sigma_1$ such that $R_1^{e_2 \cap \sigma_2}(x_0) < \varepsilon_2$ and so on. Choose $\sigma_n \subset \sigma_{n-1}$ such that $R_1^{e_n \cap \sigma_n}(x_0) < \varepsilon_n$. Let us prove that $E = \bigcup_{n=1}^{\infty} (e_n \cap \sigma_n)$ is strongly thin at x_0.

Let δ be any neighbourhood of x_0. Then

$$R_1^{E \cap \delta}(x_0) \leq R_1^{(\bigcup_1^N e_n \cap \sigma_n) \cap \delta}(x_0) + R_1^{(\bigcup_{N+1}^{\infty} e_n \cap \sigma_n) \cap \delta}(x_0).$$

The second term to the right is majorized by $R_1^{(\bigcup_{N+1}^{\infty} e_n \cap \sigma_n)}(x_0)$, then by $\sum_{N+1}^{\infty} R_1^{e_n \cap \sigma_n}(x_0)$ which is $\leq \sum_{N+1}^{\infty} \varepsilon_n$. Choose N such that $\sum_{N+1}^{\infty} \varepsilon_n < \frac{\varepsilon}{2}$. As $\bigcup_1^N (e_n \cap \sigma_n)$ is strongly thin at x_0, there exists a neighbourhood δ of x_0 such that $R_1^{(\bigcup_1^N e_n \cap \sigma_n) \cap \delta}(x_0) < \frac{\varepsilon}{2}$. Hence $R_1^{E \cap \delta}(x_0) \leq \varepsilon$ and $\operatorname{Inf}_{\sigma} R_1^{E \cap \sigma}(x_0) \leq \varepsilon$. As ε is arbitrary, we conclude that E is strongly thin.

<u>Exercises</u>. 1) Let e_n be a sequence of strongly thin sets at x_0. Let us assume that Φ is C-closed and that there exists a countable base (ω_n) of neighbourhoods of x_0 such that $\bigcap \omega_n = \{x_0\}$. Then it is possible to choose a decreasing sequence of neighbourhoods σ_n such that $\bigcap \sigma_n = \{x_0\}$ and that $\bigcap_{n=1}^{\infty} (e_n \cup \sigma_n) - \{x_0\}$ is strongly thin at x_0.

<u>Hint</u>. For any sequence of sets e_n and δ_n, we have

$$\bigcap_{n=1}^{\infty} (e_n \cup \delta_n) \subset (\bigcap \delta_n) \cup e_1 \cup (e_2 \cap \delta_1) \cup (e_3 \cap \delta_2) \cup \ldots.$$

and apply the previous theorem.

2) Let $(\mathcal{V}_n)$ be a sequence of fine neighbourhoods of x_0. Let us assume that Φ is C-closed, that thinness implies strong thinness and that there exists a countable base $\{\omega_n\}$ of neighbourhoods of x_0 such that $\cap\omega_n = \{x_0\}$. Then there exists a decreasing sequence of neighbourhoods σ_n of x_0 $\cap\sigma_n = \{x_0\}$ such that $\cup(\mathcal{V}_n \setminus \sigma_n)$ is still a fine neighbourhood of x_0.

[This is just a translation of Ex. 1 in terms of fine neighbourhoods].

Proposition II, 10. Hyper thinness implies strong thinness and in case Φ is C-closed both are equivalent.

Proof. The first part is easy. There exists a $u \in \Phi$ such that $u(x_0)$ is finite, and $u > \lambda$ on a suitable $e \cap \sigma$. Hence

$$R_{\lambda}^{e\cap\sigma}(x_0) \leq u(x_0),$$

$$R_1^{e\cap\sigma}(x_0) \leq \frac{u(x_0)}{\lambda}$$

and therefore, $\inf_{\sigma} R_1^{e\cap\sigma}(x_0) = 0$.

Conversely, suppose e is strongly thin at x_0 and that Φ is C-closed. Since $\inf_{\sigma} R_1^{e\cap\sigma}(x_0) = 0$, there exists for every n, a neighbourhood σ_n of x_0 such that $R_1^{e\cap\sigma_n}(x_0) < \frac{1}{n^3}$. Therefore, there exists $u_n \in \Phi$ such that $u_n(x_0) < \frac{1}{n^3}$ and $u_n \geqslant 1$ on $e \cap \sigma_n$. Consider $u = \sum n\, u_n$. By hypothesis, $u \in \Phi$, $u(x_0) = \sum nu_n(x_0) < \sum \frac{1}{n^2} < +\infty$. $u(x) \geqslant n\, u_n(x) \geqslant n$ on $e\cap\sigma_n$. Given λ, consider $N \geqslant \lambda$; thus $u(x) \geqslant \lambda$ on $e\cap\sigma_N$, i.e. $u(x) \to +\infty$ $x \in e$, $x \to x_0$. Therefore, e is hyperthin at x_0.

4. Strong unthinness. For e to be unthin at $x_0 \notin e$, it is necessary and sufficient that $\operatorname{Inf}_{\sigma} R_1^{e\cap\sigma}(x_0) \geqslant 1$.

Let δ be another neighbourhood of x_0. We have, for any such δ

$$R_1^{e\cap\sigma}(x_0) \geqslant R_1^{(e\cap\sigma)\setminus\delta}(x_0).$$

Therefore, $R_1^{e\cap\sigma}(x_0) \geqslant \sup_{\delta} R_1^{(e\cap\sigma)\setminus\delta}(x_0)$.

Therefore, $\operatorname*{Inf}_{\sigma} R_1^{e\cap\sigma}(x_0) \geqslant \operatorname*{Inf}_{\sigma} (\sup_{\delta} R_1^{(e\cap\sigma)\setminus\delta}(x_0))$.

Definition II, 11. e is said to be strongly unthin at a point $x_0 \notin e$

if $$\operatorname*{Inf}_{\sigma} (\sup_{\delta} R_1^{(e\cap\sigma)\setminus\delta}(x_0)) \geqslant 1.$$

This implies unthinness.

Important Remarks

1) Let e be strongly unthin at x_0. For any σ $\sup_{\delta} R_1^{(e\cap\sigma)\setminus\delta}(x_0) \geqslant 1$.

Given a σ and a number K such that 0 K 1, there exists a suitable δ such that $R_1^{(e\cap\sigma)\setminus\delta}(x_0) > K$.

2) Conversely, given any K such that $0 < K < 1$, and any σ, if there exists a δ such that $R_1^{(e\cap\sigma)\setminus\delta}(x_0) > K$, then e is strongly unthin at x_0.

3) If for any set $A \not\ni x_0$, $\sup_{\delta} R_1^{A\setminus\delta}(x_0) = R_1^{A}(x_0)$, δ being a variable neighbourhood of x_0, then A is strongly unthin at x_0, provided A is unthin at x_0.

Theorem II, 12. e_n is a sequence of strongly unthin sets at x_0, $x_0 \notin e_n$. Assume that there exists a countable base of neighbourhoods of x_0. Then there exists a suitable decreasing sequence of neighbourhoods δ_n of x_0 such that $\cup(e_n\setminus\delta_n)$ is strongly unthin at x_0.

Proof. Let λ_n be an increasing sequence of numbers $0 < \lambda_n < 1$ such that $\lambda_n \to 1$ as $n \to +\infty$. Let ω_n be a countable decreasing base of neighbourhoods of x_0. e_1 is unthin at x_0. Therefore, we can find a neighbourhood δ_1 such that $\delta_1 \subset \omega_1$ and $R_1^{(e_1\cap\omega_1)\setminus\delta_1}(x_0) > \lambda_1$. Since e_2 is unthin at x_0, choose δ_2 such that $\delta_2 \subset \omega_2 \cap \delta_1$ and $R_1^{(e_2\cap\omega_2)\setminus\delta_2}(x_0) > \lambda_2$. In general, for any n choose δ_n such that $\delta_n \subset \omega_n \cap \delta_{n-1}$ and $R_1^{(e_n\cap\omega_n)\setminus\delta_n}(x_0) > \lambda_n$.

Let $E = \cup(e_n\setminus\delta_n)$

$$(E\cap\omega_n)\setminus\delta_n \supset (e_n\cap\omega_n)\setminus\delta_n .$$

Therefore, $R_1^{(E\cap\omega_n)\setminus\delta_n}(x_0) \geqslant R_1^{(e_n\cap\omega_n)\setminus\delta_n}(x_0) \geqslant \lambda_n$.

Let $$0 < K < 1.$$

Therefore, we have for all $n \geqslant$ a suitable N,

$$R_1^{(E\cap\omega_n)\setminus\delta_n}(x_0) \geqslant K.$$

Therefore, $$\sup_{\delta} R_1^{(E\cap\omega_n)\setminus\delta}(x_0) \geqslant K.$$

Since any σ contains a ω_n where $n > N$, $\operatorname{Inf}_{\sigma} \sup_{\delta} R_1^{(E\cap\sigma)\setminus\delta}(x_0) \geqslant K$.

Therefore, $\operatorname{Inf}_{\sigma} \sup_{\delta} R_1^{(E\cap\sigma)\setminus\delta}(x_0) \geqslant 1$. Therefore, E is strongly unthin at x_0.

Remark. Suppose the unthinness is always strong at x_0 and that there exists a countable base of neighbourhoods of x_0. e_n unthin at x_0 implies x_0 is a fine limit point for e_n. The above proposition implies that if x_0 is a fine limit point for all e_n, then there exists a decreasing sequence δ_n of neighbourhoods of x_0 such that x_0 is a fine limit point of $\cup(e_n\setminus\delta_n)$.

5. Definition II, 13. It may be useful to define a negligible set e by the condition $\forall e' \subset e$, $R_1^{e'} = 0$ on $\complement e'$.

Properties: i) a finite union of such sets is again negligible

ii) a strictly polar set is negligible. Because, if $e' \subset e$, and $x \in \complement e'$, then there exists a $u \in \Phi$ such that $u = +\infty$ on e' and $u(x)$ finite; then for every $\lambda > 0$ $\lambda u = +\infty$ on e', $R_1^{e'}(x) \leqslant \lambda u(x)$. Hence, $R_1^{e'}(x) = 0$.

iii) If e is negligible, $e\setminus\{x\}$ is strongly thin at x for every x.

Suppose Φ is C-closed, then.

iv) a countable union of negligible sets is negligible.

v) a negligible set e is strictly polar. For, let $e' \subset e$ and $x \notin e'$. There exists $u_n \in \Phi$ such that $u_n \geqslant 1$ on e' and $u_n(x) < 2^{-n}$. Now, $\sum_1 u_n \in \Phi$, $\sum_1 u_n = +\infty$ on e', $\sum_1 u_n(x)$ is $< +\infty$.

CHAPTER III

General Results On Fine Limits[6]

1. With the same space Ω, basic topology $\mathcal{C}_1$ and cone Φ as in Chap. I § 1, we shall compare fine limits and ordinary $\mathcal{C}_1$-limits. We shall start with a few easy remarks.

a) Let e be unthin at x_0. Then x_0 is a fine limit point for e and conversely.

b) Let Ω' be any topological space. Let f be a function from a subset E of Ω, unthin at $x_0 \notin E$, taking values in Ω'. Let $\mathcal{V}$ be any fine neighbourhood of $x_0 \in \Omega$. Suppose f has a limit l as $x \to x_0$, $x \in \mathcal{V} \cap E$. Then on E, fine $\lim\limits_{\substack{x \to x_0 \\ x \in E}} f$ exists and is equal to l.

c) Let λ be a fine cluster value[7] of f on E at x_0, E being unthin at $x_0 \notin E$. Then λ is also a $\mathcal{C}_1$-cluster value of f at x_0 on any fine neighbourhood $\mathcal{V}$.

Theorem III, 1[8]. Let f be a function on a subset E of Ω unthin at $x_0 \notin E$ and taking values in Ω', another topological space. Let us suppose that fine $\lim\limits_{\substack{x \to x_0 \\ x \in E}} f = l$ exists. Assume also that the following are true.

i) Φ is C-closed

ii) thinness implies strong thinness

iii) In Ω', there exists a countable base of neighbourhoods for every point.

Then there exists a fine neighbourhood $\mathcal{V}$ of x_0 such that $\lim\limits_{\substack{x \to x_0 \\ x \in E \cap \mathcal{V}}} f = l$

---

(6) Inspired from some points of Cartan and Doob in classical Potential Theory and from the axiomatic development of Brelot [21].

(7) λ is a fine (resp. $\mathcal{C}_1$-clustervalue) iff the inverse image by f of any neighbourhood of λ intersects any fine (resp. $\mathcal{C}_1$)-neighbourhood of x_0.

(8) Due to H. Cartan in the classical case. See Deny [3].

(i.e. f tends to 1 outside a suitable thin set).

Proof. Let (α_n) be a decreasing sequence of neighbourhoods of 1. For every n, there exists a fine neighbourhood δ_n of x_0 such that $x \in E \cap \delta_n$ implies $f(x) \in \alpha_n$.

Let $e_n = \left\{ x \in E \,\middle|\, f(x) \notin \alpha_n \right\}$. Then, since $e_n \subset \complement \delta_n$, e_n is thin at x_0 for every n.

By ii) e_n is strongly thin at x_0 for every n. Therefore, by theorem II, 9, there exist suitable neighbourhoods σ_n of x_0 such that $e = \bigcup(e_n \cap \sigma_n)$ is strongly thin at x_0. Therefore, $\complement e$ is a fine neighbourhood of x_0.

Let α be any neighbourhood of 1 in Ω'. $\alpha \supset \alpha_n$ for large n.

Let $x \in \sigma_n \cap \complement e \cap E$ which is contained in $\complement e_n$. Therefore, $f(x) \in \alpha_n$ and hence $f(x) \in \alpha$.

Therefore, $\lim\limits_{\substack{x \to x_0 \\ x \in \complement e \cap E}} f = 1.$

In case when Ω' is the extended real line, we can even prove a stronger result with the same f defined on E unthin at $x_0 \notin E$, namely

Theorem III, 2. Let f be a real valued function and let fine $\limsup\limits_{\substack{x \to x_0 \\ x \in E}} f = \lambda$. Assume still (i) (that Φ is C-closed) and (ii) (that thinness implies strong thinness). Then there exists a fine neighbourhood $\mathcal{V}$ such that $\limsup\limits_{\substack{x \to x_0 \\ x \in E \cap \mathcal{V}}} f = \lambda$.

Proof. For any fixed fine neighbourhood $\mathcal{V}$, we have,

$$\lambda = \text{fine} \limsup_{\substack{x \to x_0 \\ x \in E}} f \leq \limsup_{\substack{x \to x_0 \\ x \in E \cap \mathcal{V}}} f \quad .$$

If $\lambda = +\infty$, the theorem is proved for any $\mathcal{V}$. Therefore, assume that $\lambda < +\infty$. Let us prove that for some suitable $\mathcal{V}$

$$\limsup_{\substack{x \to x_0 \\ x \in E \cap \mathcal{V}}} f \leq \text{fine} \limsup_{\substack{x \to x_0 \\ x \in E}} f .$$

Let λ_n be a decreasing sequence of real numbers $> \lambda$ such that $\lambda_n \to \lambda$. For any n, $\sup f \leq \lambda_n$ on a suitable fine neighbourhood of x_0. Hence, $e_n = \{x \in E \mid f(x) > \lambda_n\}$ is thin at x_0 and therefore is strongly thin at x_0 $\forall$ n. Hence, we can find a suitable decreasing sequence (δ_n) of neighbourhoods of x_0 such that $e = \bigcup e_n \cap \delta_n$ is strongly thin at x_0. Therefore, $\complement e$ is a fine neighbourhood of x_0 and $\complement e \cap \delta_n \subset \complement e_n$.

Therefore, $\sup\limits_{x \in \delta_n \cap E \cap \complement e} f \leq \lambda_n$.

Therefore, $\inf\limits_{\delta} \sup\limits_{x \in \delta \cap E \cap \complement e} f \leq \lambda_n \quad \forall\ n$.

Therefore, $\inf\limits_{\delta} \sup\limits_{x \in \delta \cap E \cap \complement e} f \leq \lambda$

i.e. $\limsup\limits_{\substack{x \to x_0 \\ x \in \complement e \cap E}} f \leq \lambda$.

Hence the equality.

Remarks. i) Let $\mathcal{V}$ be the fine neighbourhood $\complement e$ in the above theorem. Then for any fine neighbourhood $\mathcal{V}_1 \subset \mathcal{V}$, $\limsup\limits_{\substack{x \to x_0 \\ x \in \mathcal{V}_1 \cap E}} f = \limsup\limits_{\substack{x \to x_0 \\ x \in \mathcal{V} \cap E}} f = \lambda$.

Proof. $\limsup\limits_{\substack{x \to x_0 \\ x \in \mathcal{V}_1 \cap E}} f \leq \limsup\limits_{\substack{x \to x_0 \\ x \in \mathcal{V} \cap E}} f = \lambda$.

But $\limsup\limits_{\substack{x \to x_0 \\ x \in \mathcal{V}_1 \cap E}} f \geq \text{fine} \limsup\limits_{\substack{x \to x_0 \\ x \in E}} f = \lambda$.

Hence the result.

ii) Under the same hypotheses i), ii) as in theorems III, 1 and III, 2, let f_n be a sequence of functions from E to $\mathbb{R}$. Then there exists a common fine neighbourhood $\mathcal{V}$ for which the given properties hold.

Proof. Denote by $\mathcal{V}_n$, a fine neighbourhood of x_0 such that the property of theorems III, 1 or III,2 holds for f_n. We can find a suitable decreasing sequence δ_n of neighbourhoods of x_0 such that $e = \bigcup(\complement \mathcal{V}_n \cap \delta_n)$ is thin at x_0. $\complement e$ is a fine neighbourhood of x_0 and $\complement e \cap \delta_n \subset \mathcal{V}_n$ for every n. Hence $\complement e$ has the required property.

2. Theorem III, 3[9]. Let λ be a fine cluster value of f on an unthin set E at $x_0 \notin E$, where f is a function from $E \subset \Omega$ to another topological space Ω'. Then there exists an unthin set e such that $f \to \lambda$, $x \in e$, $x \to x_0$, if we assume at least the following

i) x_0 has a countable base of neighbourhoods in Ω

ii) at x_0, unthinness implies strong unthinness

iii) Ω' has a countable base of neighbourhoods at every point.

Proof. Let (α_n) be a decreasing sequence of neighbourhoods of λ in Ω'.

Let $$e_n = f^{-1}(\alpha_n).$$

Then e_n must meet any fine neighbourhood of x_0 i.e. is unthin at x_0 and therefore, by hypothesis is strongly unthin at x_0. Thanks to Theorem II, 12 we can find a decreasing sequence δ_n of neighbourhoods of x_0 such that

$$e = \bigcup(e_n \setminus \delta_n) \text{ is strongly unthin at } x_0.$$

If σ is a neighbourhood of λ, let us choose $\alpha_n \subset \sigma$. On e_p, $p \geqslant n$, $f(x) \in \alpha_p \subset \sigma$ and therefore also on $\bigcup_{p=n}^{\infty}(e_p \setminus \delta_p)$ and thus on $\delta_n \cap e$. Hence $f \to \lambda$, $x \in e$, $x \to x_0$.

---

(9) Inspired by a similar result of Doob at the Martin boundary in the classical Potential Theory.

CHAPTER IV

Quasi-topological Notions[10]

1.   In classical potential theory, properties of Lusin-type are known for a long time (H. Cartan [1]) i.e., in $\mathbb{R}^3$, any potential has a continuous restriction on a suitable set whose complementary set has a capacity arbitrarily small. And this idea was used successfully by Choquet in potential theory with kernels (Choquet [4], see also Brelot [20]). On the other hand, a theorem of Choquet [6] for instance in $\mathbb{R}^3$ says[11] that the points of $\complement e$ where e is thin may be embedded in an open set ω such that $\omega \cap e$ has an arbitrarily small capacity. And such result appears too, to be a key in the refined potential theory.

      Actually both questions are connected and are closely connected with the fine topology. That will appear in the following axiomatic introductory study which is to be completed by recent researches of Fuglede [1, 2, 3].

2.   We start from a topological space Ω and introduce on it another and finer topology, that we call fine topology (fine, finely are used for the corresponding notions). The fine closure of a set e will be denoted by $\tilde{e}$.

Definition IV, 1. Any extended real-valued set function p defined on the class of all subsets of Ω, non-negative, increasing, vanishing on the void set is called a weight[12].

      p is said to be:

fine if $p(\tilde{e}) = p(e)$, $\forall e$ ($\tilde{e}$, the fine closure of e),

countably sub-additive (c.s.a) if

    $p(\cup e_n) \le \sum p(e_n)$ for every sequence e_n of sets,

continuous to the right (c.r) if

---

(10) The corresponding lecture was made from a mimeographed course in Paris (1963 - '64) and is completed here chiefly thanks to the paper (Brelot [32])

(11) Choquet considered also the points x of e where e is thin (See Chap V § 4 and Chap VI) but the corresponding similar result for these points was well-known. We need here only the property of $\complement e$ which uses only the concept of thinness of a set e at a point of $\complement e$.

(12) This last condition is equivalent to the existence of quasi-open sets or of quasi-continuous functions.

$$p(e) = \inf_{\omega \text{ open} \supset e} p(\omega) \quad .$$

For any given weight p, we say that set ω is p-quasi-open, if given $\varepsilon > 0$, there exists an open set $\omega' \supset \omega$ such that $p(\omega' - \omega) < \varepsilon$. We say a set is p-quasi-closed if its complement if p-<u>quasi-open</u>.

We say that a function $f: \Omega \to \Omega'$ is p-<u>quasi-continuous</u> if given any $\varepsilon > 0$ there exists a set $\alpha \subset \Omega$ such that $p(\alpha) < \varepsilon$ and $f \big| \complement \alpha$ is continuous.

We define the notion of p-quasi-semi-continuity in a similar way for any real-valued function.

We say that a property holds p-<u>quasi-everywhere</u> (p q.e) if it holds except on a set e with $p(e) = 0$.

When we are dealing with a fixed weight p, we write "quasi-open" etc. instead of "p-quasi-open" etc.

<u>First properties</u>:

<u>Proposition IV.2</u>. a) If α is quasi-closed (resp. open) χ_α is quasi-upper-semi-continuous (resp. quasi-lower-semi-continuous).

<u>Proof</u>. Consider β closed, $\beta \subset \alpha$, (α quasi-closed) such that $p(\alpha - \beta) < \varepsilon$. χ_β is upper semi-continuous; $\chi_\alpha \big| \complement(\alpha - \beta)$ is equal to χ_β on $\complement(\alpha - \beta)$. Hence, χ_α is quasi-upper-semi-continuous.

b) The converse is true if p is c.r.

<u>Proof</u>. Consider e open with $p(e) < \varepsilon$ such that $\chi_\alpha \big| \complement e$ is upper-semi-continuous. Then $\chi_{\alpha - e}$ is upper-semi-continuous; hence $\alpha - e$ is closed.

c) Suppose p is c.r; if f is quasi-continuous, $f^{-1}(\omega')$ is quasi-open for any open ω'; if a real f is quasi-upper-semi-continuous $\{x \mid f < \lambda\}$ is quasi-open for any λ.

<u>Proof</u>. As for the first property, let us consider e open such that $p(e) < \varepsilon$ and $f \big| \complement e$ is continuous; then $f^{-1}(\omega') \cap \complement e$ is open on $\complement e$ and $(f^{-1}(\omega') \cap \complement e) \cup e$ is open; hence $f^{-1}(\omega')$ is contained in this open set with a difference of weight $\leq p(e)$.

Similar argument for the second part.

d) Suppose μ is c.s.a; if, for $f: \Omega \to \Omega'$, Ω' with a countable base, $f^{-1}(\omega')$ is quasi-open for any ω' open, then f is quasi-continuous. If for a real f, the set $\{x \mid f(x) < \lambda\}$ is quasi-open for every λ, then f is quasi-upper-semi-continuous.

<u>Proof</u>. As for the first property, let ω_n be a base of open sets of Ω'. Let $e_n \subset \Omega$, open such that $f^{-1}(\omega_n) \subset e_n$ and $\mu(e_n - f^{-1}(\omega_n)) < \frac{\varepsilon}{2^n}$. Consider $A = \bigcup (e_n - f^{-1}(\omega_n))$. Then $\mu(A) < \varepsilon$. It is sufficient to see that for any n, $f^{-1}(\omega_n) \cap \complement A$ is open on $\complement A$; that is a consequence of $f^{-1}(\omega_n) \cap \complement A = e_n \cap \complement A$.

Similar argument for the second property, by using instead of ω_n, the set $\{x \mid f < \lambda_n\}$ with a dense set of λ_n.

Let us emphasize on the fact that for a c.r. weight, the definitions of quasi-semi continuity remain invariant by imposing α of the definition to be open as other authors do. Hence the interest of the following remark.

__Remark__. One may define __outer weight__ μ^* as $\mu^*(e) = \inf_{\omega \text{ open} \supset e} \mu(\omega) \geq \mu(e)$

It is always c.r. It is fine, c.s.a if μ is. Note that $(\mu^*)^* = \mu^*$.

__Classical example in__ $\mathbb{R}^n (n \geq 3)$: -

The previous fine topology will be the so called classical fine topology. The outer Newtonian capacity is a weight which is c.r, c.s.a and fine. As we shall see later, the latter property is not at all as elementary as the former ones; let us recall that the sets of weight-capacity zero are all polar sets. See more developments in Chap. VI, VII, VIII.

3. __Comparison of quasi-continuity and fine continuity__.

Let us now give first an easy theorem which was known more or less in the classical potential theory (See for ex. Deny-Lions [1]).

__Theorem IV, 3.__ __If the weight__ μ __is fine, any__ μ __-quasi continuous function__ $f : \Omega \to \Omega'$ __is fine continuous__ μ __-quasi-everywhere; any__ μ __-quasi-semi-continuous function__ (__upper or lower__) __is fine semi-continuous__ μ __-quasi-everywhere__.

We choose a set α_n such that $f \mid \complement \alpha_n$ is continuous with $\mu(\alpha_n) < \frac{1}{n}$. Then $\mu(\tilde{\alpha}_n) < \frac{1}{n}$ and $\mu(\cap \tilde{\alpha}_n) = 0$. If $x \in \cap \tilde{\alpha}_n$, there exists a n_0 such that $x \notin \tilde{\alpha}_{n_0}$; $\complement \tilde{\alpha}_{n_0}$ is fine open and $f \mid \complement \tilde{\alpha}_{n_0}$ is continuous at x and therefore, is fine continuous at x. Hence, f is fine continuous at x in Ω and therefore is fine-continuous μ -quasi-everywhere in Ω. Similar proof for semi-continuity.

__Corollary.__ If μ is fine, for any quasi-closed set α, $\mu(\tilde{\alpha} - \alpha) = 0$. Similar result for quasi-open sets.

__Proof.__ χ_α is quasi-upper semi-continuous (see Prop IV, 2(a)); hence it is fine upper semi-continuous except on a set e of weight zero. Now, at any point x

of $\tilde{\alpha} - \alpha$, χ_α is zero but has a fine lim sup equal to 1, therefore is not fine upper semi continuous at x. Hence $\tilde{\alpha} - \alpha \subset e$.

Exercise: Give a direct proof of this corollary; when Ω' has acountable base and $\wp$ is c.r. and c.s.a deduce theorem IV, 3.

4. The Choquet property.

Definition IV, 4. For a fixed $\wp$, the Choquet property reads:

Any fine open set is $\wp$-quasi-open (or any fine closed set is $\wp$-quasi-closed).

Equivalent form (α). Given any e and any $\varepsilon > 0$, the fine exterior of e may be embedded in an open set ω such that $\wp(\omega \cap e) < \varepsilon$.

Note that this property implies:

(β) Any real function with values 0 or 1, which is fine u.s.c[13] must be quasi-u.s.c. Conversely, if $\wp$ is c.r (β) implies the Choquet property and hence is another equivalent form.

A weight possessing the Choquet property is said to be "of Choquet type" (briefly c.t)

Proposition IV, 5. If $\wp$ is C.t. any fine closed set is the union of a F_σ set and of a set of weight zero; if $\wp$ is also c.r, any fine Borel set is the union of a Borel set and of a set of weight zero.

Proof. The first part is easy. As for the second part, the fine Borel sets form by definition a σ-algebra which is the smallest σ-algebra containing all the fine-closed sets. The unions of a Borel set and of a set of weight zero also form a σ-algebra (consider any countable intersection and the complementary set of such a union). This second σ-algebra contains any fine closed set and hence the fine Borel σ-algebra.

Proposition IV, 6. If $\wp$ is fine and is of Choquet type, then given any set e and any $\varepsilon > 0$, e is a union of e_1 and e_2 where $\wp(\bar{e}_1) \leq \wp(e)$ and $\wp(e_2) < \varepsilon$.

---

(13) Or even whose restriction to the complement of a set of weight zero is fine u.s.c.

Proof. Same proof as in the Choquet original paper. $\tilde{e}$ contains e_0 closed such that $\mathcal{P}(\tilde{e} - e_0) < \varepsilon$ and $\mathcal{P}(e_0) \leq \mathcal{P}(\tilde{e}) = \mathcal{P}(e)$. Then we take for e_1 the set $e \cap e_0$ and for e_2 the set $e - e_0$.

5. Equivalence of quasi-continuity and fine continuity quasi-everywhere.

Theorem IV, 7. Suppose $\mathcal{P}$ is c.s.a and is C.t. Any function $f: \Omega \to \Omega'$, Ω' with countable base, fine continuous $\mathcal{P}$-quasi-everywhere (or only whose restriction on a set E with $\mathcal{P}(\complement E) = 0$ is fine continuous) is $\mathcal{P}$-quasi-continuous; any real f fine semi-continuous (u. or l.) $\mathcal{P}$-quasi-everywhere (or only whose similar restriction is fine s.c) is $\mathcal{P}$-quasi-semi-continuous (u or l).

Suppose first that f is fine continuous. Consider a base ω_n of open sets of Ω', $e_n = f^{-1}(\omega_n)$ and an open set $e'_n \supset e_n$ with $\mathcal{P}(e'_n - e_n) < \frac{\varepsilon}{2^n}$. Denote $E = \bigcup(e'_n - e_n)$; then $\mathcal{P}(E) < \varepsilon$. Let us prove that $f \mid \complement E$ is continuous. By considering the inverse images of the open sets of Ω', it is sufficient to see that $e_n \cap \complement E$ is open on $\complement E$. That is a consequence of the fact that $e_n \cap \complement E = e'_n \cap \complement E$.*(14)

Suppose now that for a set α such that $\mathcal{P}(\alpha) = 0$, $f \mid \complement\alpha$ is fine continuous on $\complement\alpha$. We may apply the previous result to f on $\complement\alpha$ with the same $\mathcal{P}$ and the induced topologies. Then f is quasi-continuous on $\Omega \setminus \alpha$, therefore on Ω. Similar argument in case of f fine u.s.c (or l.s.c) by considering the sets $\{x \mid f < \lambda_n\}$, $\{\lambda_n\}$ dense, e'_n open $\supset e_n$ such that $\mathcal{P}(e'_n - e_n) < \frac{\varepsilon}{2^n}$ etc.

Final Equivalence properties. From theorems IV, 3 and IV, 7 and definition IV, 4 (equivalence β), we conclude.

A) Theorem IV, 8. If $\mathcal{P}$ is fine, c.s.a and C.t

a) for any $f: \Omega \to \Omega'$ (Ω' with countable base), f quasi-continuous $\Longleftrightarrow$ f fine continuous q.e

b) for any real f

f quasi (u or l) s.c. $\Longleftrightarrow$ f fine (u or l) s.c, q.e

---

(14)* Argument inspired by Doob. In the classical example of n° 2, we get for any Newtonian Potential, a property of quasi-continuity due to H. Cartan.

B) Theorem IV, 9. If $\mathcal{P}$ is c.r and c.s.a, the Choquet property is equivalent to: for any real f, f fine u.s.c implies f quasi u.s.c; and if moreover $\mathcal{P}$ is fine, the Choquet property is equivalent to the identity of "fine u.s.c q.e" and "quasi u.s.c".

6. Case of a fine topology defined by a cone of functions.

We consider now a cone Φ as in Chap I, the corresponding thinness and the fine topology.

We may complete in this case, the research of sufficient conditions for the Choquet property.

Theorem IV, 10. Suppose $\mathcal{P}$ is c.s.a, c.r and Ω with a countable base. Then if the inf (the lower envelope) of any subfamily of Φ is quasi-upper semi-continuous, the Choquet property for $\mathcal{P}$ is satisfied.

Consider a set e, a base $\{\omega_n\}$ of open sets of Ω and $e_n = \{x \mid R_1^{e\cap\omega_n}(x) < 1\}$. According to Chap I, the set E of points of $\complement e$ where e is thin is $\bigcup(e_n \cap \omega_n)$. As $R_1^{e\cap\omega_n}$ is (because of the hypothesis) is quasi-upper-semi continuous, there exists an open set α_n such that $\mathcal{P}(\alpha_n) < \frac{\varepsilon}{2^n}$ and $R_1^{e\cap\omega_n} \mid \complement\alpha_n$ is upper-semi-continuous.

As $R_1^{e\cap\omega_n} \geqslant 1$ on $e\cap\omega_n$, the inequality holds on $\overline{e\cap\omega_n - \alpha_n} \subset \complement e_n$. Hence, $\complement\overline{(e\cap\omega_n - \alpha_n)} \supset e_n$.

Consider $A_n = \complement\overline{(e\cap\omega_n - \alpha_n)} \cap \omega_n \supset e_n \cap \omega_n$; $\bigcup A_n$ is open and contains E.

Now $A_n \cap e \subset \alpha_n$ or $\complement A_n \cup \complement e \supset \complement\alpha_n$ because $\complement A_n \cup \complement e$ contains $((e\cap\omega_n) - \alpha_n) \cup \complement(e\cap\omega_n)$ which contains $\complement\alpha_n$.

Hence $\bigcup A_n \cap e \subset \bigcup\alpha_n$ and $\mathcal{P}(\bigcup A_n \cap e) \leq \varepsilon$.

7. Examples of weight. From a weight $\mathcal{P}$ and an increasing real function $L(x) \geqslant 0$ $(x \geqslant 0)$, one gets a new weight $L(\mathcal{P}(e))$ whose properties may be deduced from those of $\mathcal{P}$ and L.

Let us study the cone Φ and the useful weight $\mathscr{P}(e) = R^e_\varphi(x_0)$ (see Chap II) where $x_0 \in \Omega$ and $\varphi \geqslant 0$ are fixed. If φ is lower-semi-continuous, $\mathscr{P}$ is fine (as a consequence of prop II, 5); if φ is finite continuous > 0, $\mathscr{P}$ is c.r (same proof as in the axiomatic theory of harmonic functions Brelot [20], see theorem 23, p.122); $\mathscr{P}$ is c.s.a, if Φ is C-closed.

Remark. With $\mathscr{P}(e) = R^e_1(x_0)$, any quasi-closed set α is thin at any $x \notin e$ and for any x_0 contains a closed set α_0 such that $\alpha - \alpha_0 \not\ni x_0$ and is thin at x_0.

It is easy by using in α a closed set α_0 such that $R_1^{\alpha-\alpha_0}(x_0) < 1$.

Weight $R^e_{f_0}(x_0)$ and the corresponding Choquet property.

Proposition IV, 11. Suppose that for $f_0 \geqslant 0$ on Ω, $R^e_{f_0}(x_0)$ is a c.r. weight. Then for any function $f(0 \leq f \leq f_0)$ which is quasi-upper-semi-continuous (in particular fine upper-semi-continuous quasi-everywhere when the weight is c.s.a and C.t) $\inf_\varphi R_{f-\varphi}(x_0) = 0$ ($\varphi \geqslant 0$, upper-semi-continuous, $\leq f$; $f - \varphi$ is given the value zero when $f = \varphi = +\infty$).

Proof. There exists an open set ω such that $R^\omega_{f_0}(x_0) < \varepsilon$ and $f \mid \complement\omega$ is upper-semi-continuous. This function on $\complement\omega$, continued by 0 outside becomes f_1 upper-semi-continuous on Ω with $R_{f-f_1}(x_0) \leq R^\omega_{f_0}(x_0) < \varepsilon$.

Theorem IV, 12. Let f_0 be a finite continuous function > 0 on Ω. We suppose for any fine upper-semi-continuous function $f(0 \leq f \leq f_0)$, that $\inf_\varphi R_{f-\varphi}(x_0) = 0$ (φ u.s.c, $0 \leq \varphi \leq f$). Then $R^e_{f_0}(x_0)$ has the Choquet property.

Proof. Consider a fine closed set E; $(f_0)_E$ defined as $f_0 \cdot \chi_E$ is fine upper-semi-continuous; if $\varphi \geqslant 0$ is upper-semi-continuous, $\varphi \leq (f_0)_E$, the set $\alpha_0 = \left\{x \mid \varphi(x) \geqslant \frac{f_0}{2}\right\}$ is closed and is contained in E. On $E - \alpha_0$, $(f_0)_E - \varphi = f_0 - \varphi > \frac{f_0}{2}$. Hence, $R_{(f_0)_E - \varphi} \geqslant R^{E-\alpha_0}_{\frac{f_0}{2}} = \frac{1}{2} R^{E-\alpha_0}_{f_0}$; $R^{E-\alpha_0}_{f_0}(x_0)$ is therefore arbitrarily small for a suitable φ; i.e. $R^{E-\alpha}_{f_0}(x_0)$ is arbitrarily small for a suitable closed set $\alpha \subset E$ and that is the Choquet property.

Corollary. Equivalence property relative to the weight $R^e_{f_o}(x_o)$ (f_o finite continuous > 0) supposed to be c.s.a.

The Choquet property is equivalent to:

$$\forall f,\ f \text{ fine upper-semi-continuous, } 0 \leq f \leq f_o:\ \inf_{\varphi} R_{f-\varphi}(x_o) = 0$$

(φ variable upper-semi-continuous, $0 \leq \varphi \leq f$).

Immediate consequence of IV, 11 and 12.

This equivalence means that the approximation property of any fine upper-semi-continuous function ($\leq f_o$) by a upper-semi-continuous function, expressed by the previous equality, can be interpreted as a Choquet property for a suitable weight, actually the weight $R^e_{f_o}(x_o)$.

Exercise. Suppose Ω is locally compact with a countable base, and Φ is C-closed. Then, for any finite continuous $f_o > 0$ the Choquet property corresponding to the weight $R^e_{f_o}(x_o)$ is independant of f_o, equivalent to the local property for $R^e_1(x_o)$ and to the property: $\inf_{\varphi} R_{f-\varphi}(x_o) = 0$ (φ u.s.c, $0 \leq \varphi \leq f$), $\forall f \geq 0$ fine u.s.c and locally bounded.

8. f_o-vanishing families of sets, ($f_o \geq 0$) (See Brelot [27])

Consider the general Ω and Φ of Chap. I and denote

$$\hat{f}(x) = \liminf_{y \to x} f(y).$$

Definition IV, 13. A family of sets e_i is said to be f_o-vanishing if

$$\widehat{\inf_i R^{e_i}_{f_o}} = 0.$$

It is equivalent to : $\widehat{\inf_i \hat{R}^{e_i}_{f_o}} = 0.$

In any open α, the weaker condition implies that $\forall \varepsilon$, there exists x_1 where $\inf_i \hat{R}^{e_i}_{f_o}(x_1) < \varepsilon$, then j such that $\hat{R}^{e_j}_{f_o}(x_1) < 2\varepsilon$, then $x_2 \in \alpha$ such that $R^{e_j}_{f_o}(x_2) < 3\varepsilon$. The set where $\inf_i R^{e_i}_{f_o}$ is $< 3\varepsilon$ is dense; hence the f_o-vanishing property.

Application. The Choquet property for the weights $R^e_{f_o}(x_0)$ or only for $\hat{R}^e_{f_o}(x_0)$ (for any fixed $f_0 \geqslant 0$ and any fixed x_0) implies, for any fixed set e and the family of the open sets ω_i containing the fine exterior points of e, that $\{e \cap \omega_i\}$ is f_0-vanishing.

Exercise: Discuss additional hypothesis in order that $\inf_i R^{e_i}_{f_o}(x_0) = 0$ (for one f_0 and one x_0) be equivalent to the f_0-vanishing property of $\{e_i\}$.

9. Indications on Fuglede's theory [2,3]

Fuglede considers a general capacity which is exactly a weight, but c.r, c.s.a and with the property $\mathcal{P}(\bigcup \alpha_n) = \sup_n \mathcal{P}(\alpha_n)$ for any increasing sequence of sets α_n. He proves first that, if Ω has a countable base of open sets any (non-void) family of quasi-closed sets e_i contains a countable sub-family whose intersection is contained in any e_i, upto a set α_i of weight zero. He also studies in our frame with a cone Φ, the consequences of the following axioms.

1) The considered weight is fine
2) The set of points $x \in E$ where $E - \{x\}$ is thin has weight zero
3) If $\mathcal{P}(E) = 0$, $E - \{x\}$ is thin $\forall x$
4) Every function of Φ is quasi-continuous.

Note that axiom 2 in $\mathbb{R}^n$-classical potential theory (see n° 2, and Chap. VI, Φ being the set of the non-negative hyperharmonic functions) is a key theorem as well as the Choquet property; but for a Greenian-space (see Chap VI), with the corresponding theory, this property 2 is not true unless the non-polar points are taken out. In axiomatic theories of harmonic functions issued from the classical one, this axiom 2 (even without non-polar points) is not supposed to be satisfied (and the same for the Choquet property).

Therefore, we had to introduce first weaker conditions and discuss the Choquet property. But the rather strong Fuglede's axioms have interesting consequences; one gets the Choquet property, the existence of a smaller-fine closed support for a measure which does not change the sets of weight zero (extension of a Getoor theorem) etc. In a paper to appear, Fuglede [4] enlarged his theory.

CHAPTER V

Weak Thinness[15]

1. Notations. Let f be any function from a topological space Ω to $\overline{R}$. We already defined (Chap. IV, § 8)

$$\hat{f}(x) = \liminf_{y \to x} f(y)$$

$$= \sup_{\sigma} (\inf_{y \in \sigma} f(y)) \quad (\sigma \text{ a neighbourhood of } x).$$

We complete with

$$\check{f}(x) = \liminf_{\substack{y \to x \\ y \neq x}} f(y) = \sup_{\sigma} (\inf_{\substack{y \in \sigma \\ y \neq x}} f(y))$$

which means $+\infty$ at an isolated point. It is known that $\hat{f}$ is always lower semi-continuous and that $\hat{f} = \inf(\check{f}, f)$. If f is l.s.c, then $\hat{f} = f$.

We start again from the basic notions of Chap I and we shall study a notion which is useful for a better understanding of the regular boundary points in the Dirichlet problem and of the fundamental convergence theorem in Potential theory.

Definition V, 1. A set e is said to be weakly thin at the point x_0 (x_0 may or may not belong to e) if

$$\inf_{\sigma} \hat{R}_1^{e \cap \sigma}(x_0) < 1 \quad (\sigma, \text{ a neighbourhood of } x_0).$$

Remarks. 1) This weak thinness of e implies the same for any $e' \subset e$.

2) As $\hat{f} \leq f$, thinness of e at $x_0 \notin e$ implies weak thinness at x_0.

3) A point x_0 is weakly thin at x_0 iff $\hat{R}_1^{x_o}(x_o) < 1$ or equivalently iff $\check{R}_1^{x_o}(x_o) < 1$.

Examples. At any point x, Ω is not weakly thin, $\emptyset$ is weakly thin; no isolated point x_0 is weakly thin at x_0.

(15) See Brelot [23]

2. Proposition V, 2 (Ramaswamy). If for a point $x_0 \in e$, $e-\{x_0\}$ is weakly thin at x_0, but not thin, then e is weakly thin at x_0.

Proof. For any neighbourhood σ of x_0, $x_0 \in \widetilde{(e-\{x_0\})\cap\sigma}$, because $(e-\{x_0\})\cap\sigma$ is unthin at x_0. Therefore, $e\cap\sigma \subset \widetilde{(e-\{x_0\})\cap\sigma}$. As $R_1^{\alpha} = R_1^{\tilde{\alpha}}$ (Proposition II, 5),

$$R_1^{(e-\{x_0\})\cap\sigma} = R_1^{\widetilde{(e-\{x_0\})\cap\sigma}} \geq R_1^{e\cap\sigma}$$

and the same is true for the $\hat{R}_1$. The hypothesis implies that $\hat{R}_1^{(e-\{x_0\})\cap\sigma}(x_0) < 1$ for a suitable neighbourhood σ of x_0. Hence, $\hat{R}_1^{e\cap\sigma}(x_0) < 1$.

Application. If $\Omega-\{x_0\}$ is weakly thin at x_0, then it is thin at x_0.

Definition V, 3. A point x_0 is said to be singular if $\Omega-\{x_0\}$ is thin (equivalently weakly thin) at x_0.

Any isolated point is a singular point.

Lemma V, 4. Suppose x_0 is negligible then, for any set e, $\check{R}_1^e(x_0) = \check{R}_1^{e-\{x_0\}}(x_0)$.

Proof. $$R_1^e \leq R_1^{e-\{x_0\}} + R_1^{x_0} .$$

As $$R_1^{x_0}(x) = 0 \ \forall \ x \neq x_0,$$

$$R_1^e(x) \leq R_1^{e-\{x_0\}}(x) \ \forall \ x \neq x_0.$$

Hence the equality and $\check{R}_1^e(x_0) = \check{R}_1^{e-\{x_0\}}(x_0)$.

Lemma V, 5. If x_0 is not singular, then for any set e, $\check{R}_1^e(x_0) \leq R_1^e(x_0)$.

Proof. If $u \in \Phi$, $u \geq 1$ on e, then $R_1^e \leq u$, $\check{R}_1^e(x_0) \leq \check{u}(x_0)$. As $\Omega-\{x_0\}$ is unthin at x_0, $\check{u}(x_0) = u(x_0)$ (Chap. 1, § 1). Therefore, $\check{R}_1^e(x_0) \leq u(x_0)$ and hence $\check{R}_1^e(x_0) \leq R_1^e(x_0)$.

Theorem V, 6. Let x_0 be negligible and not singular. Then, for any set $e \ni x_0$, if $e-\{x_0\}$ is weakly thin at x_0, then e is weakly thin at x_0.

Proof. For a suitable neighbourhood σ of x_0,

$$\hat{R}_1^{(e-\{x_0\})\cap\sigma}(x_0) < 1.$$

But $$\check{R}_1^{(e-\{x_0\})\cap\sigma}(x_0) \leq R_1^{(e-\{x_0\})\cap\sigma}(x_0)$$

because of the last lemma.

As $$\hat{f} = \inf(f, \check{f}),$$

$$\hat{R}_1^{(e-\{x_0\})\cap\sigma}(x_0) = \check{R}_1^{(e-\{x_0\})\cap\sigma}(x_0)$$

According to lemma V, 4, this is equal to $\check{R}_1^{e\cap\sigma}(x_0)$.

Finally, $\hat{R}_1^{e\cap\sigma}(x_0) \leq \check{R}_1^{e\cap\sigma}(x_0) < 1$. Therefore, e is weakly thin at x_0.

Remark. Simple examples of Ramaswamy, show that the previous condition "x_0, not singular" is also necessary and that the union of two weakly thin sets, at x_0 may not be weakly thin; and also that all points can be singular.

3. Convergence or lower envelope theorem.

Theorem V, 7. (Brelot [23]). Let $\{u_i\}$ be a family of functions of the cone Φ. Then the set $\{x \mid \widehat{\inf u_i}(x) < \inf u_i(x)\}$ is a countable union of sets that are weakly thin at every point.

Proof. Let $e_n = \{x \mid \widehat{\inf u_i} < \inf(n, (\inf u_i) - \frac{1}{n})\}$. Then $\{x \mid \widehat{\inf u_i}(x) < \inf u_i(x)\} = \bigcup e_n$.

Let us prove that each e_n is weakly thin at everypoint of the space. Let x_0 be any point of Ω. First, let us assume that $\widehat{\inf u_i}(x_0) < +\infty$. Since $\widehat{\inf u_i}$ is lower semi-continuous at x_0, there exists a suitable neighbourhood σ of x_0 such that

$$\widehat{\inf u_i}(x) > \widehat{\inf u_i}(x_0) - \frac{1}{2n} \quad \forall x \in \sigma.$$

On e_n, $\widehat{\inf u_i}(x) < \inf u_i(x) - \frac{1}{n}$.

Therefore, on $e_n \cap \sigma$, $\inf u_i > \widehat{\inf u_i} + \frac{1}{2n}$.

Let $$k_n = \widehat{\inf u_i}(x_0) + \frac{1}{2n}.$$

Therefore, $$\frac{u_i}{k_n} > 1 \text{ on } e_n \cap \sigma,$$

$$\frac{u_i}{k_n} \geq R_1^{e_n \cap \sigma},$$

$$\operatorname{Inf} \frac{u_i}{k_n} \geqslant R_1^{e_n \cap \sigma}$$

$$\frac{\widehat{\operatorname{Inf} u_i}}{k_n} \geqslant \widehat{R}_1^{e_n \cap \sigma}$$

$$\widehat{R}_1^{e_n \cap \sigma}(x_o) \leqslant \frac{\widehat{\operatorname{Inf} u_i}(x_o)}{k_n} < 1.$$

Therefore, $\operatorname{Inf}_{\sigma} \widehat{R}_1^{e_n \cap \sigma}(x_o) < 1$. Therefore, e_n is weakly thin at x_o.

Suppose, $\widehat{\inf u_i}(x_o) = +\infty$. Then, given any n, there exists a neighbourhood σ' of x_o such that $\widehat{\inf u_i}(x) \geqslant n \; \forall x \in \sigma'$. Therefore $e_n \cap \sigma' = \emptyset$; consequently, e_n is thin and hence weakly thin at x_o.

Extension. We have not used anywhere in the above proof the lower semi-continuity of the u_i, nor the additive property of Φ. Hence the validity of the theorem with obvious definitions of reduced functions and weak thinness, applied to such an extended Φ.

A countable union of sets which are weakly thin at any point is called semi-polar(16). Theorem V, 7 means that the "exceptional set" defined by the inequality is semi-polar.

4. Thinness of a set e at $x_o \in e$

Definition V, 8. e will be said to be thin at $x_o \in e$ if $e - \{x_o\}$ is thin at x_o and e weakly thin at x_o.

Definition V, 9. The base B_e of a set e is the set of the points where e is not thin.

Obviously $e = B_e \, U_e$.

These notions will be important in particular theories.

Remark. In important applications $R_\varphi^e = \widehat{R}_\varphi^e$ on $\complement e$ (for any $\varphi \geqslant 0$). Hence thinness $\Longleftrightarrow$ weak thinness everywhere.

As a consequence of theorem V, 6, we get

---

(16) In classical Green space and in axiomatic theories with enough axioms, polar sets are characterized by the property of thinness at any point. (See the following general definition). Here the analogous concept with the weak thinness is the semi-polar one.

Proposition V, 10. Suppose (i) no point is singular, (ii) for any x_0, "x_0 negligible" is equivalent to " x_0 weakly thin at x_0 i.e. $\hat{R}_1^{\{x_0\}}(x_0) < 1$, then the thinness of e at x_0 is equivalent to "$e - \{x_0\}$ thin at x_0 and $\{x_0\}$ negligible" and the fine closure of any e is the union of B_e and of the set of the negligible points of e which are fine isolated on e(17).

The hypothesis of the previous remark and proposition are satisfied in the classical theory, which we shall deeper.

Note. At the end of these generalities on fine topology, let us only mention that a deeper study leads to the consideration and comparison of various properties like connectedness and local connectedness, Lindelöf property, normality, paracompactness, countability of semi-polar sets in various hypothesis (classical case, further axiomatics.) We will refer only to Doob [8], Fuglede [4][5] and Berg (still unpublished researches) and, in the probabilistic frame, to P. A. Meyer [2].

---

(17) In similar researches, some authors avoid the existence of non-negligible points (i.e. not polar, if Φ is C-closed), because the base B_e then becomes fine perfect (i.e. without fine isolated points on B_e). That allows a simpler topological language.

But in the classical potential theory (See Chap. VI), it is natural to introduce the point at ∞ of $\mathbb{R}^n$ (even in the domain of study) in order to avoid old distinctions between interior and exterior Dirichlet problems; and for $n \geq 3$, this point is not polar in a Greenian domain containg it. In $\mathbb{R}^1$ all points are non-polar and in the axiomatic frame of harmonic functions, it is easy (by using $\mathbb{R}^1$ and $\mathbb{R}^3$ spaces) to build spaces with non-countable sets of polar and non-polar points, and a B_e which is fine closed but non-fine-perfect. (See Chap. VI).

CHAPTER VI

Notions in Classical Potential Theory[18]

1. Before applying the previous general developments to the classical potential theory, we recall some notions (see for example Brelot [25]). We shall consider even the so called $\mathcal{E}$-spaces and Green spaces, but one can think chiefly of bounded domains of $\mathbb{R}^n$.

Harmonic functions.

In an open set $\Omega \subset \mathbb{R}^n$ $(n \geq 2)$, a finite real continuous function u is said to be harmonic if, for any open ball $B_{x_0}^r$ (centre x_0, radius r) whose closure is in Ω, $u(x_0)$ is equal to the mean of u on $\partial B_{x_0}^r$ i.e. $u(x_0) = \int u \, d\sigma_{x_0}^r$, $\sigma_{x_0}^r$ being the unitary positive uniform (i.e. invariant by rotation or proportional to the area) measure on $\partial B_{x_0}^r$. That implies

a) Impossibility of a maximum or minimum at a point without constance in a neighbourhood.

b) In a domain, if u is harmonic ≥ 0, then $u > 0$ everywhere or $u = 0$ everywhere.

c) In the Alexandroff compactification $\overline{\mathbb{R}^n}$ of the space $\mathbb{R}^n$ (with point $\mathcal{A}$ at ∞) and any open set $\omega \subset \mathbb{R}^n$, $\liminf u \geq 0$ at any boundary point for u harmonic in ω implies $u \geq 0$ in ω.

The Poisson Integral for a ball $B_{y_0}^R$ is

$$I_f^{B_{y_0}^R}(y) = \int R^{n-2} \frac{R^2 - |y-y_0|^2}{|x-y|^n} f(x) \, d\sigma_{y_0}^R(x)$$

where f is summable - $d\sigma_{y_0}^R$. The definition of the integrals $\overline{I}_f$, $\underline{I}_f$ is

---

(18) Some points could be proved perhaps in a better way as in the axiomatic theory of harmonic functions. We only recall and complete classical results in a self-contained way.

obvious for any real f. They are $+\infty$ or $-\infty$ or else is harmonic in $B^R_{y_o}$.

The Dirichlet problem for any open set ω consists of finding a harmonic function in ω, tending to given values on $\partial\omega$ at every boundary point (in $\overline{\mathbb{R}}^n$). For a real boundary function f, there is at most one solution; there is one in the case of $B^R_{y_o}$, for f finite continuous.

As fundamental consequences of this property of I_f, let us mention

α) a local criterion of harmonicity.

u finite continuous and for every y_0, the existence of a $\varepsilon > 0$, such that $r < \varepsilon$ implies $u(y_0) = \int u \, d\sigma^r_{y_o}$.

β) Convergence property. for any increasing directed family u_i of harmonic functions on a domain ω, sup u_i is $+\infty$ or a harmonic function.

γ) The positive harmonic functions in a domain ω, equal to 1 at $x_0 \in \omega$ are equi-continuous at x_0 and also at any other point. This set of functions is therefore compact for the topology on it of the uniform convergence on every compact set of ω.

2. Hyperharmonic and superharmonic functions.

By adaptation of the famous Riesz theory,

u is said to be hyperharmonic on an open set $\omega \subset \mathbb{R}^n$ if

a) u is $> -\infty$

b) u is lower-semi-continuous

c) $u \geq I_u^{B^R_{y_o}}$ in any $B^R_{y_o} \subset \overline{B}^R_{y_o} \subset \omega$ or equivalently, $u(y_0) \geq \int u d\sigma^r_{y_o}$

for any $y_0 \in \omega$ and sufficiently small r. That implies the same conditions a,b,c of n° 1 (but only for min in a), the preservation of hyperharmonicity on an open set ω by limit relative to an increasing sequence or directed set or by changing the function u inside $B^R_{y_o}$ ($\overline{B}^R_{y_o} \subset \omega$) by $\overline{I}_u^{B^R_{y_0}}$.

We say u is hypoharmonic if $-$ u is hyperharmonic.

In a domain ω, a hyperharmonic (resp. hypo) u is everywhere $+\infty$ (resp $-\infty$) or else is finite on a dense set and is locally Lebesgue summable;

in this latter case, the function is said to be superharmonic (resp. subharmonic). By using the Poisson integral, given a superharmonic function u, one may build an increasing sequence of finite continuous superharmonic functions tending to u.

Notation. (2)
$$h(r) = \frac{1}{r^{n-2}} \quad (n \geq 3)$$
$$= \log \frac{1}{r} \quad (n = 2)$$

h_{x_0} is the function $x \rightsquigarrow h(|x - x_0|)$.

Examples: a) If μ is a positive Radon measure with compact support,

(3) $U^{\mu}(x) = \int h(|x - y|)\, d\mu(y)$ is superharmonic and harmonic outside the closed support of μ.

(b) If u,v are harmonic in ω, open, $|u+ iv|$ is subharmonic; in $\mathbb{R}^2$ if f is holomorphic in ω, then $|f|$ is subharmonic, $\log |f|$ is hypoharmonic and is subharmonic in any domain where $f \not\equiv 0$.

3. Notions about the point at infinity $\mathcal{A}$ (See Brelot [6])

In an open set ω of $\overline{\mathbb{R}}^n$ containing the Alexandroff point $\mathcal{A}$, u is said to be harmonic if it is finite real continuous, harmonic outside $\mathcal{A}$ and equal at $\mathcal{A}$ to $\int u\, d\sigma^r_{y_0}$ for any $B^r_{y_0}$ such that $\complement B^r_{y_0} \subset \omega$. A real u is said to be hyperharmonic in ω if it is $> -\infty$, lower-semi-continuous, hyperharmonic outside $\mathcal{A}$ and satisfying

$$u(\mathcal{A}) \geq \int u\, d\sigma^r_{y_0}, \text{ for the same } B^r_{y_0}.$$

For the exterior (in $\overline{\mathbb{R}}^n$) of a ball $B^r_{y_0}$ (exterior containing $\mathcal{A}$) denoted $B'^r_{y_0}$, we introduce

(4)
$$I_f^{B'^r_{y_0}}(y) = \int R^{n-1} \left[\frac{|y-y_0|^2 - R^2}{|x - y|^n} - \frac{1}{|y-y_0|^{n-2}} + \frac{1}{R^{n-2}} \right] \times f(x)\, d\sigma^R_{y_0}(x)$$

which becomes simpler in case $n = 2$. That plays a role like the previous I_f (about the Dirichlet problem, the definition of hyperharmonic functions, the changing in $B'^r_{y_0}$ of a hyperharmonic u by $I_u^{B'^r_{y_0}}$).

Easy extension of the previous theory.

4. $\mathcal{E}$-spaces (See Brelot-Choquet [1]).

A $\mathcal{E}$-space is a Hausdorff connected space with the following property: For every point x, there exists an open neighbourhood $\mathcal{V}_x$ and a homeomorphic mapping $y \rightsquigarrow \pi_x(y)$ of $\mathcal{V}_x$ onto an open set of $\overline{\mathbb{R}}^n$ satisfying the following condition: for any two x_1, x_2, the correspondance between $z_1 \in \pi_{x_1}(\mathcal{V}_{x_1} \cap \mathcal{V}_{x_2})$ and $z_2 \in \pi_{x_2}(\mathcal{V}_{x_1} \cap \mathcal{V}_{x_2})$ defined by $\pi_{x_1}^{-1}(z_1) = \pi_{x_2}^{-1}(z_2)$ or equivalently, the correspondance expressed by the mapping $\pi_{x_2} \circ \pi_{x_1}^{-1}$ is a) isometric (then the points with image at $\mathcal{A}$ are called points at infinity) or b) in case $n = 2$, conformal (directly or not). Such a space is locally connected, locally compact and metrizable. Any connected subspace is a $\mathcal{E}$-space. The points at infinity are isolated. A particular case is any domain of $\overline{\mathbb{R}}^n$.

The definition of harmonic, hyperharmonic, superharmonic functions is local by the same property on an image $\mathcal{V}'_x$. The properties a,b,c of § 1 (property c understood with an Alexandroff compactification of $\mathcal{E}$) remain valid. We call abstract potential (shortly potential) on the space or on an open subset any non-negative superharmonic function whose greatest harmonic minorant (which alway exists) is zero. The inf and sum of two potentials are potentials(19). If v is a potential (in the given space Ω) harmonic in ω and w hyperharmonic $\geqslant 0$ on ω such that $\liminf\,(w - v) \geqslant 0$ on ω at every boundary point of ω in the space, then $w - v \geqslant 0$ on ω.

In fact, $\inf\,[(w - v), 0]$ on ω continued by zero becomes hyperharmonic in Ω ; it is $\geqslant -v$, therefore majorises the least harmonic majorant of $-v$, i.e. zero. By taking w equal to a constant, we get the Maria-Frostman principle when v is finite continuous, namely $\sup_{\Omega} v = \sup_S v$ (S, support of v, complementary set of the greatest open set where v is harmonic). The result

---

(19) If P_1, P_2 are two potentials, h the g.h.m of $P_1 + P_2$, then $h \leqslant P_1 + P_2 \Rightarrow h - P_1 \leqslant P_2 \Rightarrow h - P_1 \leqslant 0$ (since $h - P_1$ is subharmonic and P_2 is a potential) $\Rightarrow h \leqslant P_1 \Rightarrow h \leqslant 0 \Rightarrow h = 0$.

holds for any v and may be proved by approximation. See also Th. VIII, 4.

Polar sets. We shall take in a $\mathcal{E}$-space Ω, for functions of Φ (Chap I), the non-negative superharmonic functions and the function $+\infty$. Same convention for a connected subspace. Note the countable addivity in Φ. Then a set on a domain ω will be polar in ω (according to Chap I) if there exists on ω, a positive[20] superharmonic function which is $+\infty$ (at least) on e. And one may build one which is finite at any given $x \notin e$. In an open set a subset is said to be polar if it is so in every component. That means it is strictly polar relative to the cone of non-negative hyperharmonic functions on the open set. A set e is said to be locally polar on an open set ω, if $\forall\ x \in \omega$, there exists an open neighbourhood ω_0 of x where $e \cap \omega_0$ is polar (and it is equivalent to consider only the neighbourhoods of the points of e)[21]. We say quasi-everywhere for "except on a locally polar set. A point x (as identified to $\{x\}$) is not locally polar, if and only if it is a point at infinity in case $n \geqslant 3$. When e is locally polar and closed in ω, any superharmonic function on $\omega - e$, locally bounded from below, may be continued on e in a unique way, as a superharmonic function on ω. Note also that if ω is connected, $\omega - e$ is connected.

5. Green space and Green functions.

A $\mathcal{E}$-space Ω is called a Green space when there exists on Ω a potential > 0 or equivalently a non-constant positive superharmonic function. A domain ω in a Green space is again a Green space; a domain in a $\mathcal{E}$ - but a

---

(20) Without the restriction "positive", we get a notion which is equivalent to the local property (see further) in our classical case but perhaps not in general axiomatics when there is no potential > 0.

(21) This needs the study of the neighbourhood in $\overline{\mathbb{R}^n}$ of $\mathcal{A}$ as well as of any other point.

non-Green space is a Green space iff $\Omega - \omega$ is not locally polar. On a Green space, locally polar sets are polar and even strictly polar or equivalently negligible. On a Green space all potentials which are harmonic outside a point x_0 are proportional. One is determined and called Green function $G_{x_0}^{\Omega}$ or G_{x_0} by the following condition: Close to x_0, it must be a function $f(x)$ such that $f(\pi_{x_0}^{-1}(x'))$ where $x' = \pi_{x_0}(x)$ is upto a harmonic function equal to $h(|x' - x'_0|)$ if $x'_0 \neq \mathcal{A}$ (Alexandroff point of $\mathbb{R}^n$) and to $h(|x'|)$ if $x'_0 = \mathcal{A}$.

Observe that, for x non-polar, $y \rightsquigarrow G_x(y)$ is bounded continuous and recall the important symmetry property: $G_x(y) = G_y(x)$. Hence the notation $G(x,y)$.

Examples: $\mathbb{R}^n$, $\overline{\mathbb{R}}^n$, Riemann surfaces are $\mathcal{E}$-spaces.

$\mathbb{R}^n (n \geq 3)$ and hyperbolic Riemann surfaces are Green spaces.

$\mathbb{R}^2$ and parabolic Riemann surfaces are not Green spaces.

In $\overline{\mathbb{R}}^n (n \geq 3)$ $\complement \overline{B}_0^R$ (containing $\mathcal{A}$) is a Green space and

$$G_{\mathcal{A}}(x) = \frac{1}{R^{n-2}} - \frac{1}{|x|^{n-2}} .$$

On a $\mathcal{E}$-space Ω, let us consider an open set Ω_0; every component of Ω_0 is Greenian (i.e. a Green space) unless Ω is not Greenian. In the first case, we say Ω_0 is Greenian and we may define a Green function $G_x^{\Omega_0}$ as equal to G_x^{ω} in the connected component ω of Ω_0, containing x and to zero elsewhere in Ω_0. And the symmetry holds.

Remark. In a Green space Ω, G_{x_0} has a global peak point at x_0 (See Chap 1 § 1).

In fact $G_{x_0}(x) \leq G_{x_0}(x_0)$ (Maria-Frostman principle § 4). The equality at $x \neq x_0$ is not possible (if $G_{x_0}(x_0)$ is finite, G_{x_0} would be constant). Consider in a given neighbourhood δ_0 of x_0 another neighbourhood δ contained in $\mathcal{V}_{x_0}$ (see § 4) and whose image is a ball $B_{y_0}^r$ or a $B'^r_{y_o}$. We change G_{x_0} in δ by the Poisson integral and get a superharmonic function majorized by its sup on $\partial\delta$ which is $< G_{x_0}(x_0)$. Therefore $\sup_{\complement \delta_0} G_{x_0} < G_{x_0}(x_0)$.

6. First general key results and problems.

α) The Great Convergence theorem (Brelot-Cartan)[22]

On an open set Ω of a $\mathcal{G}$-space, let us consider a decreasing directed set of superharmonic functions which are locally uniformly bounded from below. Then $\widehat{\inf u_i}$ is superharmonic and equal quasi-everwhere to $\inf u_i$ (local property). This is more precise than the theorem V, 7.

β) Approximation Lemma.

On a Green space Ω, any finite continuous function on a compact set K may be approached upto ε, by a difference of two finite continuous non-negative superharmonic functions or even potentials.

Separation property for finite continuous superharmonic functions $u \geqslant 0$ may be deduced from G_{x_0}; $|u_1 - u_2|$ and the constants are differences of such u. Then a Stone's theorem gives the approximation result. With potentials, we build a fixed finite continuous potential $V > 0$ and we work with the quotients by V of all differences of finite continuous non-negative potentials.

γ) The Dirichlet problem (Perron-Wiener-Brelot).

Consider first in a $\mathcal{G}$-space, a relatively compact open set Ω, with $\complement\Omega$ not locally polar and a real-valued function f on $\partial\Omega$. The hyperharmonic functions u on Ω satisfying "lim inf at every boundary point x, is $\geqslant f(x)$" have a lower envelope $\overline{H}_f^{\Omega}$ (shortly $\overline{H}_f$) which is in every component $+\infty$, $-\infty$ or else is harmonic. We denote $\underline{H}_f = -\overline{H}_{(-f)}$; always $\underline{H}_f \leqslant \overline{H}_f$. In case $\underline{H}_f = \overline{H}_f$, finite everywhere, f is said to be resolutive and the common envelopt "generalised solution H_f".

If ω open $\subset \Omega$, F the function $\overline{H}_f^{\Omega}$ continued by f, then $\overline{H}_F^{\omega} = \overline{H}_f^{\Omega}$ in ω.

When f is finite continuous, f is resolutive (Wiener) (use the previous approximation lemma) and (5) $H_f^{\Omega}(x) = \int f(y)\, d\rho_x^{\Omega}(y)$ where $d\rho_x^{\Omega}$ is

---

(22) See Brelot [1], Cartan [1], Brelot [6].

a positive unitary measure on $\partial\Omega$ (harmonic measure at x). In general, $\overline{H}_f(x) = \int f\, d\rho_x^{\Omega}$ and f is resolutive iff f is $d\rho_x^{\Omega}$-summable for all x or only one in every component of Ω. (Resolutivity theorem of Brelot [3]).

Remark. If v superharmonic has a harmonic minorant, the greatest one is the limit of $H_v^{\Omega_n}$ ($\overline{\Omega}_n \subset \Omega$, $\Omega_n \uparrow$, $\cup \Omega_n = \Omega$). The case where it is 0 characterises v as a potential.

As a consequence of the additivity of $f \leadsto H_f$ for resolutive f, we get the additivity of the greatest harmonic minorant relative to v (superharmonic $\geqslant 0$). This can also be proved directly by making use of the addivity of the potentials (See footnote 19).

Regular boundary points.

A boundary point X is called regular if for any finite continuous f, $H_f \longrightarrow f(X)$ as $x \longrightarrow X$. A necessary and sufficient condition is the existence in $\Omega \cap \delta$ (δ, a neighbourhood of X) of a strictly positive superharmonic function tending to 0 at X (local condition) or when Ω is connected, the condition $G_x(y) \longrightarrow 0$, $y \longrightarrow X$ (x fixed $\in \Omega$) (Bouligand's form of criterion).

Note that $X \in \partial\Omega$ is irregular if and only if it is irregular for a component of Ω and that for any upper bounded f and any regular X,

$$\limsup_{\substack{x \in \Omega \\ x \longrightarrow X}} \overline{H}_f(x) \leqslant \limsup_{\substack{y \in \partial\omega \\ y \longrightarrow X}} f(y) \quad .$$

Example of regular X. Case when in $\mathcal{V}_X$, there is a cone (vertex X' not at infinity) whose non-empty interior, close to X' is outside the image of $\mathcal{V}_X \cap \Omega$. Hence the construction of Ω' ($\overline{\Omega}' \subset \Omega$, Ω' containing a given compact K) with all boundary points regular.

A point at infinity on $\partial\Omega$ is regular when $n \geqslant 3$.

As an application of the convergence theorem, the set of irregular points of a domain (therefore of any Ω) is locally polar (Kellog-Evans). We consider $\Omega_n \uparrow$, $\overline{\Omega}_n \subset \Omega$, $\cup\Omega_n = \Omega$, with regular boundary points. $G_{x_0}^{\Omega_n}$ continued by 0 is subharmonic outside x_0. The limit $G_{x_0}^{\Omega}$ continued by 0, then regularised by lim sup becomes subharmonic (outside x_0) and differs from 0 only on a locally polar set which is the set of irregular points.

Extension. Consider a $\mathcal{E}$-space Ω_0 and denote Ω_0' the space Ω_0 when compact, if not, the compactified space with Alexandroff point at infinity $\mathcal{A}$. Consider in Ω_0, a Greenian open set Ω. One may develop a similar Dirichlet problem for Ω with $\partial\Omega$ defined on Ω_0'. Same definition of envelopes, same resolutivity theorems. For f given on Ω_0, non-compact we denote by f_* the continuation by 0 at $\mathcal{A}$. Hence notations like $H_{f_*}^{\Omega_0}$. Same remark on the addivity of the g.h.m. Same comparison with a subspace. Same definition and properties of regularity of $X \in \partial\Omega$, $X \neq \mathcal{A}$. Ω is said regular if $\mathcal{A} \notin \partial\Omega$ and all boundary points are regular. Same property that the irregular points $\neq \mathcal{A}$ from a locally polar set.

δ) Dirichlet problem for compact sets. (Keldych-Lavrentiff-Brelot)

Given a compact K in a Green space Ω, consider a finite continuous f on ∂K. The superharmonic v on an open neighbourhood of K, satisfying $\liminf_{x \in \complement K,\, x \to X} v(x) \geq f(X)$ $(X \in \partial K)$ have a lower envelope $\overline{K}_f$ on K. Similar $\underline{K}_f$ with subharmonic functions. Both envelopes are equal to the limit K_f according to the filtre of neighbourhoods ω of K, of H_F^{ω}, F being any finite continuation of f. A point $X \in \partial K$ is called stable if $\forall$ f, $K_f(X) = f(X)$. The stability of every $X \in \partial K$ is equivalent to the uniform approximation of any previous f on ∂K by means of harmonic functions defined on neighbourhoods of K. See more details in Brelot [7].

7. The Riesz representation.

For a Green space Ω (or a Greenian open set of a $\mathcal{E}$-space), for any non-negative measure μ on Ω, $\int G^{\Omega}(x,y)\, d\mu(y)$ is hyperharmonic; in every component it is $+\infty$ or an abstract potential (as defined § 4). Conversely, an abstract potential on Ω has the representation $\int G^{\Omega}(x,y)\, d\mu(y)$ (called also G^{Ω}-potential of μ)(23) with a unique $\mu \geq 0$.

A superharmonic function u on Ω with a harmonic minorant has the representation

(6) $u(x) = \int G^{\Omega}(x,y)\, d\mu(y) + u^*(x)$ where u^* is the greatest harmonic minorant in Ω and μ a unique non-negative measure (associated measure); has a local character, because its restriction on any $\omega \subset \Omega$ is the associated measure for ω.

8. Order on superharmonic functions.

We define a "specific order" by $u_1 \succ u_2 \Leftrightarrow u_1 = u_2 +$ superharmonic function ≥ 0.

If u_1 = pot of μ_1 + a harmonic function h_1

u_2 = pot of μ_2 + a harmonic function h_2,

the condition $u_1 \succ u_2$ is equivalent to $\begin{cases} \mu_1 \succ \mu_2 \\ h_1 \geq h_2 \end{cases}$ i.e. $\mu_1 - \mu_2$ is a positive measure.

It is obvious for potentials and easy to complete. Moreover, the set of the superharmonic functions ≥ 0 is a complete lattice for the specific order. In the case of potentials, the weaker condition $u_1 \geq u_2$ implies $\mu_1(\Omega) \geq \mu_2(\Omega)$. We may see that by introducing an open $\Omega' \subset \overline{\Omega'} \subset \Omega$ and $R_1^{\Omega'}$ which is a potential U^{ν} of a measure ν. For a suitable Ω', $\int U^{\nu} d\mu_1$ is arbitrarily close to $\mu_1(\Omega)$ finite or not and

$$\int U^{\nu} d\mu_1 = \int U^{\mu_1} d\nu \leq \int U^{\mu_2} d\nu = \int U^{\nu} d\mu_2 \leq \mu_2(\Omega).$$

9. Various complements in a Green space.

α) On a compact polar set e, there exists a measure $\nu \geq 0$ whose potential is $+\infty$ on e and finite on $\complement e$ (Evans). Extension by Deny [2], Choquet [3] to $\mathcal{G}_\delta$ sets.

---

(23) We will not use, unless specification the word of potential corresponding to general measure or for a difference of two previous (non-negative superharmonic) potentials.

β) If u is superharmonic (with associated measure μ), finite on a polar set e, the outer μ-measure of e is zero. e is contained in a Borel polar set and in the Borel set where u is finite. Therefore, we may suppose e Borel, then even compact and that u is the potential of a measure μ. We introduce $K_n = \{x \in e, u \leq n\}$ and the corresponding ν (Evans) of potential v, $\int v\, d\mu = \int u\, d\nu$ finite, hence $\mu(K_n) = 0$, $\mu(e) = 0$.

γ) We shall need also the existence of a finite continuous potential $U > 0$ with the following property: for every $x_0 \in \Omega$, $U = u_1 + u_2$ where u_1 is a finite continuous potential and u_2, a finite continuous function $\geqslant 0$, for which x_0 is a global peak point (See Chap 1 § 1).

We cover Ω by a sequence of open sets ω_i ($\bar{\omega}_i$ compact containing no point at infinity, but contained in some $\mathcal{V}_x$) and ω^i containing respectively the points at infinity. For every ω_i, we consider the finite continuous potential V_i, corresponding to the measure, which is zero outside ω_i and is defined on ω_i by its image equal to the Lebesgue measure on the image ω_i'. For $x_0 \in \omega_i$, let us consider a ω_i^1 whose image is a ball of centre x_0', then V_i^* deduced from V_i by changing it on ω_i^1 by the function which is on the image, the Poisson integral for the ball. $V_i - V_i^*$ is equal to zero outside ω_i^1 and equal in ω_i^1 to a function which becomes on the image the $G^{(\omega_i^1)}$-potential of the Lebesgue measure; we see that $V_i - V_i^*$ satisfies the conditions (a) (b). Consider a ω^i and the point at infinity X^i that it contains. In case $n > 2$, we introduce V^i equal to the Green function with its pole at X^i, in case $n = 2$, it is easy to define on a suitable $\omega_1^i \subset \omega^i$, a measure whose G^{Ω}-potential is finite continuous and whose $G^{\omega_1^i}$- potential continued by zero is continuous and satisfies the a,b conditions for X^i(by working on the image, and using the inversion to come back to the neighbourhood of the origin. We take for V^i, the G^{Ω}-potential of this measure). Finally we consider $\sum \lambda_i V_i + \sum \lambda^i W^i$ with suitable positive λ_i, λ^i (such that $\sum \lambda_i \sup V_i$, $\sum \lambda^i \sup V^i$ are finite). This is a wanted U.

Easy extension to an open subspace of Ω.

δ) Improvement of the approximation lemma § 6, β.

Given f finite continuous ≥ 0 on Ω, equal to zero outside a compact set $K \subset \Omega$, and in Ω a compact $K_1 \supset \overset{\circ}{K}_1 \supset K$, there exist G^{Ω}-potentials finite continuous V_1, V_2 with $V_1 - V_2 \geq 0$, $V_1 - V_2 = 0$ on $\complement K_1$, $|V_1 - V_2 - f| < \varepsilon$ on K.

Let us introduce Ω_1 (open), K_2 (compact) such that $\overset{\circ}{K}_1 \supset \bar{\Omega}_1 \supset \Omega_1 \supset K_2 \supset \overset{\circ}{K}_2 \supset K$ such that the boundary points of $\Omega_1 - K_2$ are regular.

We apply the previous lemma § 6, β, to get G^{Ω_1}-potentials u_1, u_2 such that $|u_1 - u_2 - f| < \varepsilon$ on K. We introduce $|u_1 - u_2|^+$ which is a difference of similar potentials, u_1', u_2' and change u_1', u_2' in $\Omega_1 - K_2$ by the solution of the Dirichlet problem for the values zero on $\partial\Omega_1$ and u_1', resp u_2' on ∂K_2. The new functions will be continued by zero and become finite continuous everywhere, but subharmonic outside K_2. Therefore, these functions U_1, U_2 are in Ω, locally, therefore globally, differences of finite continuous G^{Ω}-potentials (of measures on K_1) $\omega_1 - \omega_1'$ and $\omega_2 - \omega_2'$ upto harmonic functions on Ω which are actually 0. We see that $\omega_1 + \omega_2'$ and $\omega_2 + \omega_1'$ satisfy the wanted condition for V_1, V_2.

10. Reduced functions and balayage (24)

We take again in a Green space Ω the non-negative hyperharmonic functions as functions of Φ (Chap. I). For any $\varphi \geqslant 0$, and $e \subset \Omega$ note a series of properties.

a) $\hat{R}^e_\varphi$ is hyperharmonic and when superharmonic, is harmonic outside $\bar{e}$. If $\varphi > 0$, the condition $\hat{R}^e_\varphi = 0$ at a point or everywhere or $R^e_\varphi = 0$ somewhere or only $\hat{R}^e_1 < 1$ everywhere are equivalent to the property that e is polar or also negligible.

b) $\hat{R}_\varphi(x) = \inf_{\omega \in \mathfrak{F}} \overline{\int} R_\varphi \, d\rho^\omega_x$, for the family $\mathfrak{F}$ of open neighbourhoods of x.

c) $\hat{R}^e_\varphi = R^e_\varphi$ on $\complement e$ and q.e on e.

d) $\hat{R}^e_\varphi$ is the smallest hyperharmonic function $\geqslant 0$ on Ω, $\geqslant \varphi$ q.e on e (hence does not change by altering e by a polar set). Note also $\hat{R}^e_{\hat{R}^e_\varphi} = \hat{R}^e_\varphi$.

e) If $\varphi_n \uparrow \varphi$, $\hat{R}_{\varphi_n} \uparrow \hat{R}_\varphi$; $e_n \uparrow e$, $\hat{R}^{e_n}_\varphi \uparrow \hat{R}^e_\varphi$.

f) If φ is continuous, finite or not, $\hat{R}^e_\varphi = \inf_\omega \hat{R}^\omega_\varphi$, ω open $\supset$ e upto a plar set.

---

(24) The balayage is here and further as in the paper of Brelot [8] in 1945 (but in an improved redaction) for any set e by using methods which may be extended or adapted to axiomatic theories of harmonic functions, and which are based, for a full development on the great convergence theorem. But a completely different method was given by Cartan [2] a little later (explicitly in $\mathbb{R}^n$, $n \geqslant 3$); it was based on the usage of energy and Hilbert spaces and gave rise to extensions in other directions. As for previous works on balayage theory, chiefly for compact sets by De La Vallée Poussin, Frostman etc. see the bibliography in Brelot [14].

Let us mention too, a "second type of balayage" inspired by the property (d) where the condition "q.e on e" is replaced by "except on a set whose every closed subset is polar". See the corresponding and similar developments in Brelot [8] and Cartan [2]. They seem to be less useful and are not studied here. For R^e_W and $\hat{R}^e_W$ with W superharmonic $\geqslant 0$ further properties and other proofs may be given (without the use of the Dirichlet problem and of the great convergence theorem) which are valid in a large axiomatic frame. See Boboc, Constantinescu, Cornea [2]

11. Properties of $\hat{R}^e_V$ for V superharmonic ≥ 0. Balayage (sweeping out process). Notations and language.

$\hat{R}^e_V$ is said the balayaged function (swept out function) relative to V and e, denoted also $\mathcal{B}^e_V$; if V is the G^{Ω}-potential of μ, $\mathcal{B}^e_V$ is the G^{Ω}- potential of a measure denoted b^e_{μ}, called balayaged measure (swept out measure). G^{Ω}_x is the potential of the unit Dirac measure ε_x and $\mathcal{B}^e_{G_x}$ is the G^{Ω}-potential of $b^e_{\varepsilon_x}$. Recall:

Properties of $\mathcal{B}^e_V$. a') For an open set ω, $\mathcal{B}^{\complement\omega}_V = H^{\omega}_{V_*}$ on ω where V_* is equal to V in Ω and to zero at the Alexandroff point of Ω.

b') Let us consider an open set $\omega \subset \Omega$ and a boundary point x_0 (not the Alexandroff point $\mathcal{A}$ of Ω). Denote σ and V respectively, any neighbourhood of x_0 and any finite continuous potential > 0. The regularity of x_0 for ω is equivalent to $\mathcal{B}^{\complement\omega\cap\sigma}_V(x_0) = V(x_0)$ or to $R^{\complement\omega\cap\sigma}_V(x) \to V(x_0)$, $x \in \omega$, $x \to x_0$.

α) for all σ and V

β) for one compact σ and all V

γ) for one V and all σ.

Because of (a'), the regularity implies (α); thanks to the approximation lemma, (β) implies $H^{\omega-\sigma}_{\varphi} \longrightarrow \varphi(x_0)$ at x_0 for every finite continuous φ, equal to zero at $\mathcal{A}$, therefore also without this last condition, and that means the regularity. Finally (γ) implies (α); if not, for a σ and a suitable V', $\mathcal{B}^{\complement\omega\cap\sigma}_{V'}(x_0) < V'(x_0)$. Let us choose λ such that $\lambda V(x_0)$ is strictly between these numbers. On a suitable $\sigma_1 \subset \sigma$, $\lambda V < V'$ and

$$\mathcal{B}^{\complement\omega\cap\sigma_1}_{\lambda V}(x_0) \leq \mathcal{B}^{\complement\omega\cap\sigma_1}_{V'}(x_0) \leq \mathcal{B}^{\complement\omega\cap\sigma}_{V'}(x_0) < \lambda V(x_0).$$

Hence $\mathcal{B}^{\complement\omega\cap\sigma_1}(x_0) < V(x_0)$, a contradiction.

b") For a compact k and $x_0 \in \partial k$, observe that $K_V = R^{\complement k}_V$ on k and the stability of x_0 is equivalent to

$R_v^{\complement k}(x_0) = v(x_0)\ \forall\ v$ (finite continuous potential).

c') $v \rightsquigarrow \mathcal{B}_v^e$ is <u>additive</u>. By (a') it is true for e closed, then by (§ 10, e) for e open, then for any e, for v continuous by (f), then for any v and e (thanks to a sequence of continuous $v_n \uparrow v$ and to (e)). This implies the addivity of $\mu \rightsquigarrow b_\mu^e$.

d') (7)
$$\mathcal{B}_{G_x}^e(y) = \mathcal{B}_{G_y}^e(x).$$

Thanks to (e,f), we have to consider only the case when e is closed, even compact. When x,y are on $\complement e$, the result comes from the property that, for any potential $u \geqslant 0$, finite continuous on ∂e, the greatest harmonic minorant on $\complement e$ is equal to R_u^e. Then $G_x^\Omega = G_x^{\complement e} + R_{G_x}^e$ on $\complement e$ and the symmetry of the Green functions implies the wanted one. If x and y are polar (and perhaps are on e), we add to $\complement e$ open neighbourhoods of x and y, which, afterwords will be shrunk to these points. If x or y or both are not polar, and one at least is on e, we use the property of G_x (finite continuous if x non-polar) and the regularity of any non-polar boundary point. In case x and y are on e, x non-polar, y polar, we may add again to $\complement e$, a neighbourhood of y.

12. <u>Properties involving balayaged measures.</u>

e') For every superharmonic function $v \geqslant 0$ on Ω:

(8)
$$\mathcal{B}_v^e(x) = \int v \, d\, b_{\varepsilon_x}^e .$$

Hence for a potential u of μ,

(9)
$$\mathcal{B}_u^e = \int \mathcal{B}_{G_x}^e \, d\mu .$$

First, if $\varphi \geqslant 0$ finite continuous with compact support, is a difference $V_1 - V_2$ of continuous potentials, consider $\mathcal{B}_{V_1}^e(x) - \mathcal{B}_{V_2}^e(x)$, independant of the decomposition (see c') and linear (in φ). By the approximation lemma, there exists a unique Radon measure $\mu_x \geqslant 0$ on Ω, independant of

φ, V_1, V_2 such that

$$\mathcal{B}^{e}_{V_1}(x) - \mathcal{B}^{e}_{V_2}(x) = \int (V_1 - V_2)\, d\mu_x .$$

If v is a potential, finite continuous, consider $v_1 = v, v_2 = \mathcal{B}_v^{\complement\Omega_n}$ ($\Omega_n \uparrow$ with regular boundary points, $\bar{\Omega}_n \subset \Omega$, $\cup \Omega_n = \Omega$) which tends to zero, we get $\mathcal{B}^{e}_{v}(x) = \int v\, d\mu_x$ and that holds for every potential, then every superharmonic $\geqslant 0$ (thanks to increasing sequence of suitable functions).

Finally, as an application, we get $\mathcal{B}^{e}_{G_x}(y) = \mathcal{B}^{e}_{G_y}(x) = \int G_y(z)\, d\mu_x(z)$ and conclude $\mu_x = b^{e}_{\varepsilon_x}$.

Application 1 (to harmonic measure). The general formula implies that for a relatively compact open set ω, we have

$$H_f^{\omega} = \int f\, d\, b^{\complement\omega}_{\varepsilon_x}\ , \text{ thanks to } H_v^{\omega} = \mathcal{B}_v^{\complement\omega} \text{ for a potential } v$$

and to the approximation lemma.

Hence $b^{\complement\omega}_{\varepsilon_x}$ is identical to the harmonic measure for ω and x. And we may deduce for any open ω that $b^{\complement\omega}_{\varepsilon_x}$ is the restriction on $\partial\omega \cap \Omega$ of the harmonic measure at x_0 for ω [25].

Application 2. Consider μ, ν of potentials u, v and for some e, the balayaged measure μ', ν' and balayaged potentials u', v', then

$$(10) \qquad \int u\, d\nu' = \int v'\, d\mu = \int u'\, d\nu = \int v\, d\mu' .$$

Thanks to (7),

f') b^{e}_{μ} is the integral of the family of measures $b^{e}_{\varepsilon_x}$ with respect to μ i.e. for every φ finite continuous with compact support, $\int \varphi\, d\, b^{e}_{\varepsilon_x}$ is $d\mu$-summable and

$$(11) \qquad \int \varphi\, d\, b^{e}_{\mu} = \int \left(\int \varphi\, d\, b^{e}_{\varepsilon_x} \right) d\mu(x).$$

---

(25) Historically, the integral expression of H_f^{ω} in $\mathbb{R}^3$ was given (De La Vallée Poussin [1]) by using the measure obtained by another balayage theory.

More generally, we will prove that if ψ is $d\,b^e_\mu$ summable, $\int \psi \, d\,b^e_\mu$ has a meaning a.e - $d\mu$ and

$$(12) \qquad \int \psi \, d\,b^e_\mu = \int \left(\int \psi(y) \, d\,b^e_{\varepsilon_x}(y) \right) d\mu(x)$$

that implies for a Borel set α

$$(13) \qquad b^e_\mu(\alpha) = \int b^e_{\varepsilon_x}(\alpha) \, d\mu(x).$$

It is sufficient to consider φ, $\psi \geqslant 0$ and the case of a potential ψ is known thanks to the approximation lemma (§ 9 δ), there exists $\theta_n \geqslant 0$ with compact support, difference of finite continuous potentials, with compact supports, converging uniformly to φ and majorised by a constant and even by a fixed λG_{x_0}, which is $d\mu$ and $d\,b^e_\mu$-summable.

$$\int \theta_n \, d\,b^e_\mu \longrightarrow \int \varphi \, d\,b^e_\mu .$$

Thanks to (10), the first member is the $d\mu$-integral of

$$\int \theta_n \, d\,b^e_{\varepsilon_x} \text{ which is majorised by } \int \lambda G_{x_0} \, d\,b^e_{\varepsilon_x} \leq \lambda G_{x_0} \text{ and}$$

tends to $\int \varphi \, d\,b^e_{\varepsilon_x}$. Hence, the general formula (11). Then, we introduce for a general ψ, $\psi_n \downarrow$ l.s.c $\geqslant \psi$ and $\psi'_n \uparrow$, u.s.c $\geqslant 0$ and $\leq \psi$ such that $\int \psi_n \, d\,b^e_\mu$ and $\int \psi'_n \, d\,b^e_\mu$ tend to $\int \psi \, d\,b^e_\mu$. Then

$$\int \psi \, d\,b^e_\mu = \int \left(\int \lim \psi_n \, d\,b^e_{\varepsilon_x} \right) d\mu$$

$$= \int \left(\int \lim \psi'_n \, d\,b^e_{\varepsilon_x} \right) d\mu$$

hence $\overline{\int} \psi \, d\,b^e_{\varepsilon_x}$ and $\underline{\int} \psi \, d\,b^e_{\varepsilon_x}$ are equal a.e $d\mu$ and we get (12).

<u>Remark</u>. For a potential u of μ, $\mathcal{B}^e_u \leq u$, hence $b^e_\mu(\Omega) \leq \mu(\Omega)$ (see end of § 8). Observe that this is a consequence of the case $\mu = \varepsilon_x$ where the corresponding property is contained in the formula (8) of (e') for $v = 1$.

CHAPTER VII

Classical Fine Topology-General Properties

1. We start from a Green space Ω and the cone Φ of the non-negative hyperharmonic functions. Let us observe first that the topology $\mathscr{E}_1$ of Ω is also $\mathscr{E}_0$ i.e. the coarsest topology for which the functions of Φ are lower semi-continuous. That comes for instance from the fact that any $x_0 \in \Omega$ is a global peak point for G_{x_0} (see Chap I, § 1 and Chap VI, § 5). Recall that polar sets, strictly polar sets, negligible sets are the same and are thin at every point.

Theorem VII, 1. The thinness of e at $x_0 \notin e$ is equivalent to the existence of a superharmonic function v in an open neighbourhood ω_0 of x_0, such that

$$v(x_0) < \liminf_{\substack{x \in e \\ x \to x_0}} v(x)$$

(according to the general meaning recalled in Chap I). This implies the local character of thinness.

It remains to see that this condition implies the thinness relative to Ω and to Φ. The measure associated to v in a neighbourhood of x_0 has a G-potential which satisfies the same inequality. It is possible to avoid this measure to get a non-negative superharmonic function on Ω, which is, close to x_0, equal to V, upto a harmonic function. (See axiomatic theories).

Fine topology in a $\mathscr{G}$-space. If we take for thinness the property of the previous theorem, the complementary sets of thin sets satisfy the neighbourhood axioms (as in any Green subspace) and define the so-called fine topology in the given space; it induces on any Green subspace, the fine topology of this one.

Remark. Any thin set e at $x_0 \notin e$ is contained in an open set $\delta \not\ni x_0$, thin at x_0.

2. We come back to a Green space Ω and try to use the developments of Chap I - V. We shall develop and complete some parts of classical potential theory, following methods of Brelot [8][26].

Theorem VII, 2. Thinness (of e at $x_o \notin e$) is always strong (and is equivalent to hyperthinness; see Chap I, n° 4, Chap II n° 3).

Proof. If u is superharmonic $\geqslant 0$, finite at x_o,

$$u(x) = \int_{\mathcal{V}-\{x_o\}} G(x,y)\, d\mu(y) + \int_{\complement\mathcal{V}} G(x,y)\, d\mu(y) + \mu(\{x_o\})\, G_{x_o}$$

+ a harmonic function, where $\mathcal{V}$ is a neighbourhood of x_o, and μ is associated to u. The second integral and the further term are finite at x_o and are continuous ($\mu\{x_o\}$ is zero if $\{x_o\}$ is polar). The first integral is arbitrarily small for a suitable $\mathcal{V}$. Then Theorem II, 8 completes the proof.

Exercise. Prove directly that thinness implies hyperthinness.

Theorem VII, 3. Unthinness (of e at $x_o \notin e$) is always strong

Proof. We have only to see that $\sup_{\delta} R_1^{A-\delta}(x_o) = R_1^A(x_o)$ ($x_o \notin A$) for a variable neighbourhood δ of x_o or only for a decreasing sequence δ_n ($\bigcap \delta_n = \{x_o\}$). Now using Chap VI, § 8,

$$R_1^{A-\delta_n}(x_o) = \hat{R}_1^{A-\delta_n}(x_o)$$

$$\hat{R}_1^{A-\delta_n} \uparrow \hat{R}_1^{A}$$

$$\hat{R}_1^{A}(x_o) = R_1^{A}(x_o) \qquad \text{(b and d)}.$$

Hence the wanted property.

Then the theorems III, 1,2,3 on fine limits are applicable.

Theorem VII, 4. No point of Ω is singular (i.e. $\Omega-\{x\}$ is always unthin at x).

Thinness of e at x_o is equivalent to:

$e-\{x_o\}$ thin and $\{x_o\} \cap e$ polar or also to the weak thinness.

---

(26) See Chiefly Brelot [8], H. Cartan [1, 2].

A criterion is $\inf_{\sigma} \hat{R}_1^{e \cap \sigma}(x_0) < 1$ (equivalently $= 0$), σ, a variable neighbourhood of x_0.

<u>Hence semi-polar $\Leftrightarrow$ polar $\Leftrightarrow$ negligible.</u>

<u>Proof.</u> The lower semi-continuity and mean inequality of superharmonic functions imply that $\Omega - \{x_0\}$ is not thin.

The further properties are consequences of Chap V, § 4.

Note that polar sets are thin at every point (and do not contain non-polar points). Thinness of e at x_0 is preserved by altering e by a polar set.

<u>Theorem VII, 5. If σ is any neighbourhood of</u> x_0 <u>and</u> v <u>any finite continuous potential</u> > 0 <u>on Ω</u>, unthinness of e at x_0 is equivalent to the condition

$$\hat{R}_v^{e \cap \sigma}(x_0) = v(x_0)$$

α) $\forall \sigma$ and $\forall$ v

or β) $\forall$ v, for one σ

or γ) $\forall \sigma$, for one v

<u>Proof.</u> If $x_0 \notin e$, $\hat{R}_v^{e \cap \sigma}(x_0) = R_v^{e \cap \sigma}(x_0)$.

The prop II, 4 and 5 show the equivalence with (γ) and also the necessity of α, β.

Suppose β ; the thinness of $e \cap \sigma$ would imply for the potential U(Chap VI, § 9γ) according to prop II, 6, $R_U^{e \cap \sigma}(x_0) < U(x_0)$. We conclude $e \cap \sigma$ and e are unthin.

Suppose $x_0 \in e$; if x_0 is non-polar, there is unthinness and (α); if x_0 is polar the proposition is equivalent to the same for $e - \{x_0\}$.

3. <u>Theorem VII, 6. There exists a finite continuous potential</u> $U > 0$ <u>such that for every set e, the set of the points where e is thin is</u>

$\{x, \hat{R}_U^e(x) < U(x)\}$.

Proof. It is sufficient to take the same U(Chap VI, § 9). The previous theorem shows that this inequality implies thinness and we have just recalled how the thinness implies the inequality.

Theorem VII, 7 (Brelot [5, 6]). The set of points of a set e where e is thin is a polar set.

Proof. It is obvious that $R_U^e = U$ on e and the convergence theorem implies $\hat{R}_U^e = R_U^e$ q.e.

Corollary. A polar set e is characterised as a set which is thin at every point, or equivalently, at every point of e, or as a set made of polar fine isolated points.

Exercise. Prove theorem VII, 7 without the use of U, with a countable base of open sets, thanks to the criterion of Th. VII, 4.

Base of a set (set of points where e is unthin, Def. V, 9).

It is the set of the fine accumulation points[27] for e and of the non-polar points of e. The fine closure $\tilde{e}$ of e is the union of B_e and of the (polar) set of the points of e where e is thin, i.e. of the polar fine isolated points of e (See Prop. V, 10).

Recall the invariance of $\hat{R}_\varphi^e$ $(\varphi \in \Phi)$ by changing e to e' upto a polar set or to $\tilde{e}$ (Prop. II, 5). Theorem VII, 6 allows then further results. If for instance, α fine closed contains e upto a polar set, denote $e' = e \cap \alpha$. Then $\alpha \supset e'$ implies $\alpha \supset \tilde{e}' \supset B_{\tilde{e}'} = B_{e'} = B_e$. Let us emphasize.

Proposition VII, 8. B_e is the smallest fine closed set containing e upto a polar set $B_{\tilde{e}} = B_{B_e} = B_e \subset \tilde{e}$. B_e is a G_δ -set; obviously $B_e \subset \bar{e}$ (ordinary closure of e).

Proposition VII, 9. A B_e set is characterised (and called a base) as a set E which is unthin at every point of E and thin at every point of $\complement E$.

Observe that $\tilde{e}$ and B_e have the same fine interior.

---

(27) i.e. any point such that every fine neighbourhood contains a point of e different from this one.

One may also use the <u>notion of kernel</u> K_e of e which will be $K_e = \complement B_{\complement e}$. It is the greatest fine open set contained in e upto a polar set (exactly the union of the fine interior of e and of the polar set of the points of $\complement e$ where $\complement e$ is thin.

Remember: fine interior of $e \subset K_e \subset$ fine interior of B_e or $\tilde{e}$; fine boundary of B_e is contained in the fine boundary of e.

§ 4. <u>First applications</u>.

<u>Lemma VII, 10.</u> If $\{e_i\}$ is a family of fine closed sets, there exists a countable subfamily e_{i_n} such that $B_{\cap e_{i_n}} = B_{\cap e_i}$.

<u>Proof</u>. In fact, using the function U of theorem VII, 6, we have for any e

$$B_e = \left\{x,\ \mathcal{B}_U^e = U\right\}.$$

As $\widehat{\inf \mathcal{B}_U^{e_i}} = \widehat{\inf \mathcal{B}_U^{e_{i_n}}}$ for a suitable countable subfamily e_{i_n} by application of a topological lemma of Choquet (see Brelot [25], Chap I), we get

$$\mathcal{B}_U^{\cap e_{i_n}} \leq \mathcal{B}_U^{e_{i_n}} \quad \forall\ n .$$

Therefore
$$\mathcal{B}_U^{\cap e_{i_n}} \leq \inf \mathcal{B}_U^{e_{i_n}} .$$

Hence
$$\mathcal{B}_U^{\cap e_{i_n}} \leq \widehat{\inf \mathcal{B}_U^{e_{i_n}}} = \widehat{\inf \mathcal{B}_U^{e_i}}$$

and therefore, $\mathcal{B}_U^{\cap e_{i_n}} \leq \mathcal{B}_U^{e_i} \quad \forall\ i.$

Hence, $$B_{\cap e_{i_n}} \subset B_{e_i} \subset e_i \quad \text{(fine closed).}$$

Therefore $B_{\cap e_{i_n}} \subset \cap e_i$. By taking the base of both members, $B_{\cap e_{i_n}} \subset B_{\cap e_i}$.

Hence, the identity.

<u>Applications: Prop VII, 10</u>. An intersection (resp. union) of fine closed (open sets) is identical to the intersection (union) of a suitable countable subfamily, upto a polar set. That is a particular case of a general study of Doob [8].

Proposition VII, 11. For any family of fine u.s.c. functions u_i, there exists a countable subfamily u_{i_n} such that $\inf u_{i_n} = \inf u_i$ quasi every where.

For any rational r, we choose a sequence u_{i_p} such that

$\left\{\inf\limits_p u_{i_p} \geqslant r\right\} = \left\{\inf\limits_i u_i \geqslant r\right\}$ upto a plar set. The union of all

functions of these sequences for all r gives a family $u_{i'_n}$ satisfying the previous equality for all rationals outside a polar set α. For any point x where $\inf\limits_i u_i < \inf\limits_n u_{i'_n}$, we consider a rational r strictly between these numbers and see that $x \in \alpha$.

Theorem VII, 12 (Getoor)[28]. If a measure μ does not charge polar sets (i.e., the outer measure is zero), there exists a smallest fine closed support E of which is even a base.

Proof. Every fine closed set is the union of a base and of a polar set and therefore is μ-measurable. Denote by $\{e_i\}$, the fine closed supports. $B_{\cap e_i} = B_{\cap e_{i_n}}$ for a suitable countable subfamily $\{e_{i_n}\}$. As $\complement(\cap e_{i_n}) = \bigcup \complement e_{i_n}$, $\cap e_{i_n}$ is a fine closed support. $B_{\cap e_{i_n}}$ differs from $\cap e_{i_n}$ by a polar set and is also a fine closed support. Now, any fine closed support e_i contains $\cap e_i$ (fine closed), then $B_{\cap e_i}$ or $B_{\cap e_{i_n}}$. Hence $B_{\cap e_i}$ is the smallest fine closed support.

Application to the Dirichlet Problem.

Theorem VII, 13. For an open set ω (resp. a compact set K) (in a Green space Ω) a point x_0 of $\partial\omega \cap \Omega$ (resp. of ∂K) is irregular, (resp. unstable) iff $\complement\omega$ (resp. $\complement K$) is thin at x_0[29].

Proof. Regularity and unthinness of $\complement\omega$ are equivalent to the condition $\hat{R}_v^{\complement\omega\cap\sigma}(x_0) = v(x_0)$ (for a σ and any v potential finite continuous (Chap VI, § 11, b' and theorem VII, 5).

Stability and unthinness of $\complement K$ are equivalent to : $R_v^{\complement K}(x_0) = v(x_0)$ for every v (See Chap VI § 11,b" and theorem VII, 5).

---

(28) See Getoor, Choquet [7], Doob [8], Brelot [30].

(29) This common criteria of irregularity and instability has suggested the introduction of thinness (See Brelot [4]).

We will use later other criteria for this identity of regularity and unthinness (Chap IX).

CHAPTER VIII

Applications to balayage, weights and capacities

1. We consider the same Green space and classical theory and first complete the balayage theory, following generally Brelot [8].

Observe that for a superharmonic function $u \geq 0$, $\mathcal{B}_u^e$ is equal to u on B_e and is preserved by changing e to $\tilde{e}$ or to B_e. Recall that $\mathcal{B}^e_{\mathcal{B}_u^e} = \mathcal{B}_u^e$.

Lemma VIII, 1. Given a measure $\mu \geq 0$ supported by B_e (i.e. such that $\mu(\complement B_e)=0$), then $b_\mu^e = \mu$.

Proof. From $u(x) = \int G(x,y)\, d\mu(y)$, we get (Chap VI, § 12), $\mathcal{B}_u^e(x) = \int \mathcal{B}^e_{G_x}(y) d\mu(y)$ where $\mathcal{B}^e_{G_x} = G_x$ on B_e. Hence $\mathcal{B}_u^e = u$.

Lemma VIII, 2. For every $\mu \geq 0$, $b_\mu^e(\complement B_e) = 0$.

Proof. This is a consequence by integration of measure of the particular case of a Dirac measure (See Chap VI, n° 12, f'). Consider such a ε_x and use again the potential U of theorem VII, 6; $\mathcal{B}_U^e(x) = \int U\, db^e_{\varepsilon_x}$, then by iteration $\mathcal{B}_U^e(x) = \int \mathcal{B}_U^e\, d\, b^e_{\varepsilon_x}$. Hence the set where $U > \mathcal{B}_U^e$ which is $\complement B_e$ has a $b^e_{\varepsilon_x}$ - measure zero.

Key Theorem VIII, 3 (Brelot [8]). For a potential u, the measure corresponding to $\mathcal{B}_u^e$ is the unique measure $\nu \geq 0$ satisfying

a) $\nu(\complement B_e) = 0$ (i.e. B_e is a support of ν)

b) Potential of $\nu = u$ on B_e (or equivalently q.e on e).

Proof. We know these properties of b_μ^e whose potential is $\mathcal{B}_u^e$ (see Lemma VIII,2). Now, for such a ν of potential v, q.e equal to u on e, therefore equal to u on B_e, $v = \mathcal{B}_v^e$ (Lemma VIII,1). Therefore, $v(x) = \int v\, d\, b^e_{\varepsilon_x} = \int u\, d\, b^e_{\varepsilon_x}$ (Lemma VIII, 2) i.e. $v = \mathcal{B}_u^e$.

First Important consequences:

a) Theorem VIII, 4. (Domination principle; strong form)(30). Given a potential

---

(30) Stronger than a similar principle of H. Cartan [2].

u of μ, suppose v superharmonic ≥ 0, is $\geq$ u q.e on a set E whose base B_E supports μ (i.e. $\mu(\complement B_E) = 0$), then $v \geq u$ everywhere.

Proof. It is actually a consequence of Lemma 1 alone, which implies $\mathcal{B}_u^E = u$. As $v \geq \mathcal{B}_u^E$, $v \geq u$.

b) Theorem VIII, 5. If μ of potential u does not charge the set E of irregular boundary points of an open set ω, $\mathcal{B}_u^{\complement\omega}$ is in ω, the greatest harmonic minorant of u and conversely[31].

Proof. If v^* is the greatest harmonic minorant of v (v superharmonic ≥ 0) on ω, we know that $(v_1 + v_2)^* = v_1^* + v_2^*$ (Chap VI, § 6, c). Therefore, consider the restrictions μ_1, μ_2 of μ on ω and on $\complement\omega$ with potentials u_1, u_2. We know that on ω $\mathcal{B}_{u_2}^{\complement\omega} = u_2$ (harmonic in ω) iff μ_2 charges only the base of $\complement\omega$ i.e. does not charge E. Then it will be sufficient to prove that $\mathcal{B}_{u_1}^{\complement\omega} = u_1^*$. Let us use ω_n open $\uparrow$ $\bar{\omega}_n \subset \omega$, $\bigcup\omega_n = \omega$, the potentials v_n and w_n of the restrictions of μ on ω_n and on $\omega\setminus\omega_n$. On ω, $u_1^* = v_n^* + w_n^*$, $w_n^* \leq w_n \to 0$. Now $v_n^* = \mathcal{B}_{v_n}^{\complement\omega}$. It is easy to see this when ω is relatively compact. In the general case, let us introduce $\Omega_p \uparrow$, $\bar{\Omega}_p \subset \Omega$, $\bigcup\Omega_p = \Omega$ and φ_n equal to v_n on $\partial\omega \cap \Omega$ and to 0 elsewhere. We may see that $H_{v_n}^{\Omega_p \cap \omega}$ tends to v_n^* (as $p \to \infty$) on ω and that $H_{\varphi_n}^{\Omega_p \cap \omega}$ tends to $H_{v_n}^{\omega}$, because of the convergence and increasing of the harmonic measure on $\partial\omega \cap \Omega$. * As $\left| H_{v_n}^{\Omega_p \cap \omega} - H_{\varphi_n}^{\Omega_p \cap \omega} \right| \leq H_{v_n}^{\Omega_p} \to 0$ (as $p \to \infty$), we get the wanted property.

$$\text{But } \mathcal{B}_{v_n}^{\complement\omega} \to \mathcal{B}_{u_1}^{\complement\omega}. \text{ Hence } u_1^* = \mathcal{B}_{u_1}^{\complement\omega}.$$

Remark. As a particular case, we may suppose that u is locally bounded. This implies μ does not charge polar sets, in particular the set of the irregular boundary points of ω.

---

(31) In the general case, the difference of $\mathcal{B}_u^{\complement\omega}$ (called best harmonic minorant) and of the greatest harmonic minorant ist equal to $\int G_x(y)\, d\mu(y)$ where G_x ($x \in \omega$) is suitably continued on $\partial\omega$ (unique continuation that gives with o on $\complement\omega$ a subharmonic function on $\Omega\setminus\{x\}$)

(Result of Brelot, for ω bounded in $\mathbb{R}^n$ by using an argument of Frostman, see ref in Brelot [25]).

c) Theorem VIII, 6. The set of boundary points in Ω where an open set ω is thin has harmonic measure zero[32] (contained in De la Vallée Poussin's work).

Proof. Considering the potential U of theorem VII, 6 and the potential $U' = \mathcal{B}_U^{\complement\omega} \leq U$, we see that $\mathcal{B}_U^{\complement\omega}$ and $\mathcal{B}_{U'}^{\complement\omega}$ are the greatest harmonic minorants of U, U' in ω, therefore equal, because $U = U'$ on ω. Hence $H^{\omega}_{\underset{*}{U}} = H^{\omega}_{\underset{*}{U'}}$, and $U = U'$ on $\partial\omega \cap \Omega$ a.e in harmonic measure.

2. Further properties of b^e_μ.

Lemma VIII, 7. If u is superharmonic $\leq \lambda$, finite, the (fine closed) set where $u = \lambda$ has a fine interior of μ-measure zero for the measure μ associated to u.

Proof. It is sufficient to prove the same for the intersection with an open relatively compact set ω_1, then for a positive u_1 in $\omega_2 \supset \overline{\omega}_1$, ω_2 relatively compact regular and even connected as Ω. Note that $u_1 = (R^{\omega_1}_{u_1})_{\omega_2}$ on ω_1. Finally, we come back to the case of a potential u tending to zero at the Alexandroff point and it is sufficient to prove that the set e where $u = \lambda = \sup_\Omega u$, has a fine interior of μ-measure zero. But now the open set α where $u > \lambda - \varepsilon$ is relatively compact in Ω and on $\partial\alpha$, $u = \lambda - \varepsilon$ except perhaps on the set of $\partial\alpha$, where α is thin. But this set has harmonic measure zero in α and this implies that $H^{\alpha}_u = \hat{R}^{\complement\alpha}_u$, is equal to $\lambda - \varepsilon$ in α. Hence $\hat{R}^{\complement e}_u = \lambda$ on e and $\hat{R}^{\complement e}_u = u$ q.e, therefore everywhere. As u is the balayaged function $\mathcal{B}^{\complement e}_u$, μ is supported by $B_{\complement e}$ and the kernel or the fine interior of e has μ-measure zero.

Theorem VIII, 8. Consider a measure μ corresponding to a potential such that $\mu(B_e) = 0$. For α polar $\subset B_e$, or for α = fine interior of B_e, α has an outer b^e_μ-measure zero.

Proof. As for the first part (α polar), we may by enlargement, suppose α borelian and it is sufficient to consider the case where α is compact and $\mu = \varepsilon_x$, $x \notin B_e$, G_x is bounded on α, therefore also $\mathcal{B}^e_{G_x}$ whose corresponding measure will be zero on α (see Chap VI, n° 9, β).

---

(32) As $b^{\complement\omega}_{\varepsilon_x}$ is the harmonic measure of ω at x (Chap VI, § 10) the result is contained in the next theorem VIII,8 (whose given proof needs this property (c) in the Lemma 7).

As for the second part, we consider again ε_x $(x \notin B_e)$ and apply the previous lemma to the space $\Omega \setminus \{x\}$ and the superharmonic function $\mathcal{B}^e_{G_x}(y) - G_x(y)$. This function is zero on B_e; therefore B_e has a fine interior of $\mathcal{B}^e_{\varepsilon_x}$-measure zero.

3. Applications

Theorem VIII, 9. Given u superharmonic $\geqslant 0$, $\mathcal{B}^e_u$ (which is equal to u on B_e) depends only on the values of u on the fine boundary of B_e. Hence $\mathcal{B}^e_u$, on the fine closure of $\complement e$, depends only on the values of u on the fine boundary of e.

Proof. We start from $\mathcal{B}^e_u(x) = \int u(y)\, db^e_{\varepsilon_x}(y)$ and the first part is obvious. Suppose $x \in \widetilde{\complement e}$; if $x \in B_e$, $\mathcal{B}^e_u(x) = u(x)$ and $x \in \tilde{e}$, therefore x is a fine boundary point of e. If $x \notin B_e$, we conclude because the fine boundary of B_e is contained in the fine boundary of e.

Corollary: Let us consider u_1, u_2 superharmonic $\geqslant 0$ and $A = \{x \mid u_1 = u_2\}$. Then $\mathcal{B}^{\complement A}_{u_1} + u_2 = \mathcal{B}^{\complement A}_{u_2} + u_1$, and on the kernel or on the fine interior of A, the measures corresponding to u_1 and u_2 are identical.

As $u_1 = u_2$ on A (fine closed), therefore on the fine boundary of A or of $\complement A$, we have $\mathcal{B}^{\complement A}_{u_1} = \mathcal{B}^{\complement A}_{u_2}$ on A. On the other hand, $\mathcal{B}^{\complement A}_{u_1} = u_1$ q.e. on A, and the same for u_2. Hence the wanted equality q.e, therefore everywhere.

If u_1, u_2 are potentials, $\complement B_{\complement A}$ or $K_{\complement A}$ has measure zero for the measures corresponding to $\mathcal{B}^{\complement A}_{u_1}$ and $\mathcal{B}^{\complement A}_{u2}$ and the equality we have proved gives the result on the measures. In the general case, we introduce an open set $\omega \subset \bar{\omega} \subset \Omega$ and the potentials $\mathcal{B}^{\omega}_{u_1} = u'_1$, $\mathcal{B}^{\omega}_{u_2} = u'_2$ (equal to u_1, u_2 on ω). On the fine interior of $\omega \cap A$, the corresponding measures are identical. Hence, the wanted result for u_1, u_2.

This is partially contained in the following result (improvement of a De La Vallée Poussin's theorem. See Brelot [13].

Theorem VIII, 10. With the same superharmonic $u_1, u_2 \geqslant 0$ and the set $A = \{x \mid u_1 = u_2\}$, let us introduce the subset A_0 where u_1, u_2 are finite and where the (fine open) set $\alpha = \{x \mid u_1 < u_2\}$ is thin. Then, on A_0, the measure μ_2 associated to u_2 majorises the measure μ_1 associated to u_1.

Proof. As previously, we may come back to the case of potentials u_1, u_2. Then, consider $\mathcal{B}_{u_1}^{\complement\alpha}$, $\mathcal{B}_{u_2}^{\complement\alpha}$ whose corresponding measures are equal to μ_1, μ_2 on A_o (contained in the fine interior of $B_{\complement\alpha}$). As $\mathcal{B}_{u_1}^{\complement\alpha}$, $\mathcal{B}_{u_2}^{\complement\alpha}$ are equal on α, because they are equal on its fine boundary, we come back to the case where $u_1 \geqslant u_2$ everywhere.

With these hypotheses, let us consider a compact set $K \subset A_o$ and an open set $\omega \supset K$ such that $\mu_1(\omega \setminus K), \mu_2(\omega \setminus K)$ are $< \varepsilon$. If h is the greatest harmonic minorant of u_2 in ω, consider $u_1 - h = u_1'$ and the potential $u_2' = u_2 - h$. The balayaged function $(\mathcal{B}_{u_2'}^{K})_\omega$ defined by its ordinary definition for the components of ω has a corresponding measure equal to μ_2 on K upto a measure of total $< \varepsilon$ (note that the plar set of points of thinness of K has μ_1 and μ_2 measure zero). By decomposition of u_1' in a potential u_1'' on ω and a harmonic function $\varphi \geqslant 0$ on ω, we get $(\mathcal{B}_{u_2'}^{K})_\omega = (\mathcal{B}_{u_1'}^{K})_\omega = (\mathcal{B}_{u_1''}^{K})_\omega + (\mathcal{B}_{\varphi}^{K})_\omega$. Therefore, the corresponding measure on K majorises μ_1, upto a measure of total $< \varepsilon$. Hence $\mu_2(K) \geqslant \mu_1(K)$.

Properties of $\mathcal{B}_u^e$ for a harmonic u. Because of additivity, it remains to study the case of u harmonic.

Theorem VIII,11. If u is harmonic $\geqslant 0$, the measure corresponding to $\mathcal{B}_u^e$ is supported by the fine boundary of B_e (therefore of e).

Proof. First, according to the last corollary or theorem, u and $\mathcal{B}_u^e$, equal on B_e have the same corresponding measures on the fine interior of B_e. If we decompose $\mathcal{B}_u^e$ as pot v + harmonic h, we get by the same balayage, $\mathcal{B}_u^e = \mathcal{B}_v^e + \mathcal{B}_h^e$. Hence $v = \mathcal{B}_v^e$ (and $h = \mathcal{B}_h^e$). The measure corresponding to v is supported by B_e. Hence, the wanted result.

Remark. We have mentioned (Chap VI § 10, footnote 24), the balayage theory of Cartan [2]. In this paper, a concept of fine convergence on the set of the measures $\geqslant 0$ (with potential $\not\equiv +\infty$) applied to the Dirac measures ε_x gives by $\varepsilon_x \Longleftrightarrow x$, a notion of convergence on the set $\mathbb{R}^n$ which is our fine convergence.

4. Examples of weights and capacities. Choquet property.

Let us recall from Choquet's theory (see Choquet [1] [5], Brelot [20]) that a general capacity (called true capacity in the previous book of Brelot [20]) is an extended real (i.e. finite or not) set function $\mathcal{C}(e)$ defined for any set e in a Hausdorff space such that i) $\mathcal{C}(e)$ is increasing, ii) for an increasing sequence e_n, $\mathcal{C}(e_n) \longrightarrow \mathcal{C}(\cup e_n)$, iii) for a decreasing sequence of compact sets e_n, $\mathcal{C}(e_n) \longrightarrow \mathcal{C}(\cap e_n)$. The capacitability of e means $\mathcal{C}(e) = \sup_{K \text{ compact} \subset e} \mathcal{C}(K)$.

Choquet proved this property for the so-called K-analytic sets (particularly the borelian sets in our theory) contained in a G_δ-set.

An example of general capacity is given starting from a real finite $\mathcal{C}(K) \geq 0$ (K compact) such that a) $\mathcal{C}(K)$ is increasing, b) $\mathcal{C}(K)$ is "continuous to the right" which in a locally compact metric space is equivalent to the previous (iii), c) $\mathcal{C}(K_1 \cup K_2) + \mathcal{C}(K_1 \cap K_2) \leq \mathcal{C}(K_1) + \mathcal{C}(K_2)$ (strong subadditivity). This $\mathcal{C}(K)$ is called a Choquet capacity.

We introduce the inner capacity $\mathcal{C}_*(e) = \sup_{\substack{K \text{ compact} \\ K \subset e}} \mathcal{C}(K)$, the outer capacity $\mathcal{C}^*(e) = \inf_{\omega \text{ open} \supset e} \mathcal{C}_*(\omega)$. Choquet proved that $\mathcal{C}^*(e)$ is a general capacity, for which capacitability means $\mathcal{C}_*(e) = \mathcal{C}^*(e)$. $\mathcal{C}^*(e)$ is countably subadditive.

In the previous classical frame with a Green space Ω and hyperharmonic non-negative functions, we shall study, in relation with Chap IV, the set functions

$A(e) = R^e_\varphi(x)$, for any fixed x, where φ is finite continuous ≥ 0

$A_m(e) = \int \hat{R}^e_\varphi(x)\, dm(x)$ (same φ) where m is a measure ≥ 0 which does not charge polar sets and either with compact support or such that there exists V superharmonic ≥ 0 satisfying $V \geq \varphi$ and $\int V\, dm < +\infty$. Example: $d\rho^{\omega_0}_{x_0}$ harmonic measure of a relatively compact ω_0 at $x_0 \in \omega_0$, $\gamma(e) = \mu(\Omega)$ where μ is the measure corresponding to $\hat{R}^e_\varphi$ (same φ).

Theorem VIII, 12. $A(e)$, $A_m(e)$ are weights which are fine, continuous to the right, countably subadditive and of Choquet type. The second one is a general capacity and if φ is moreover superharmonic, both weights are outer capacities of Choquet capacities. When e is contained in any fixed $\Omega' \subset \overline{\Omega'} \subset \Omega$, $\gamma(e)$ has the same properties as A_m in Ω'. The outer capacity corresponding to $\gamma(K)$ when $\varphi = 1$ is called outer Greenian capacity (or only classical capacity). A set e is polar when $\varphi > 0$, $m \neq 0$, iff $A_m(e) = 0$ or $A(e) = 0$ $(x \notin e)$ or $\gamma(e) = 0$ $(\bar{e} \subset \Omega)$ or classical capacity equal to zero.

Proof. a) Consider $e \to R_\varphi^e(x_o)$. We know it is a fine, c.s.a. weight. It is c.r. when it is finite, there exists a superharmonic $u > 0$ satisfying $u \geqslant \varphi$ on e; if $0 < \lambda < 1$, the set $\{x \mid u \geqslant \lambda\varphi\}$ contains a neighbourhood α of e; $u \geqslant \lambda R_\varphi^\alpha$; $u \geqslant \lambda \inf_{\omega \text{ open} \supset e} R_\varphi^\omega$; $u \geqslant \inf_\omega R_\varphi^\omega$; $R_\varphi^e(x_o) = \inf_\omega R_\varphi^\omega(x_o)$. Suppose φ is superharmonic, from $\hat{R}_\varphi^{e_n} \to \hat{R}_\varphi^{\cup e_n}$, $R_\varphi^e = \varphi$ on e, $\hat{R}_\varphi^e = R_\varphi^e$ on $\complement e$, we get $R_\varphi^{e_n}(x) \to R_\varphi^{\cup e_n}(x)$. Hence R_φ^e is a general capacity. Moreover $K \to R_\varphi^K(x)$ satisfies the strong subaddivity. $R_\varphi^{K_1 \cup K_2} + R_\varphi^{K_1 \cap K_2} \leq R_\varphi^{K_1} + R_\varphi^{K_2}$. The property is obvious for x in $K_1 \cup K_2$ and outside by the interpretation of both members as solutions of Dirichlet problems in $\complement(K_1 \cup K_2)$ with values of these members on $\partial(K_1 \cup K_2)$ and zero at infinity. We conclude that $K \to R_\varphi^K(x)$ is a Choquet capacity and thanks to the first properties, that the corresponding outer capacity is $R_\varphi^e(x)$.

b) As for $\int \hat{R}_\varphi^e \, dm$ or $\int R_\varphi^e \, dm$, it is fine, c.s.a. Let us prove the continuity to the right. When the integral is $+\infty$ or $m = 0$, it is obvious. If not, there exists a superharmonic $u \geqslant 0$, $u \geqslant \varphi$ on e and satisfying $\int u \, dm < +\infty$ (thanks to the hypothesis on m). Their lower envelope in R_φ^e. Hence $\hat{R}_\varphi^e = \widehat{\inf}\, u_n$ (for a suitable decreasing subsequence because of a topological lemma of Choquet, see Brelot [20], p. 3).

Denote open sets $\omega_n = \{x \mid u_n > \theta\varphi\}$, $0 < \theta < 1$. Then $u_n \geqslant \theta R_\varphi^{\omega_n}$.

$$\int R_\varphi^e \, dm = \int \inf u_n \, dm = \lim \int u_n \, dm$$
$$\geqslant \theta \inf_n \int R_\varphi^{\omega_n} \, dm$$
$$\geqslant \theta \inf_{\omega \supset e} \int R_\varphi^\omega \, dm .$$

Hence, $\int R_\varphi^e \, dm \geqslant \inf \int R_\varphi^\omega \, dm$. Hence the equality.

We see immediately the limit property for $e_n \uparrow$. Therefore $\int R_\varphi^e \, dm$ is a general capacity.

When φ is superharmonic, and using a), we may conclude that $\int R_\varphi^K dm$ is a Choquet capacity. The inner capacity of ω is $\sup_{K \subset \omega} \int R_\varphi^K \, dm$ or $\int R_\varphi^\omega \, dm$ and the outer capacity of e is $\inf_{\omega \text{ open} \supset e} \int R_\varphi^\omega \, dm$ or $\int R_\varphi^e \, dm$.

c) The increasing property of $e \to \hat{R}_\varphi^e$ $(e \subset \Omega')$ implies the same for $\gamma(e)$ (see Chap VI, § 8 end). Let us use a similar argument and a potential W equal to 1 on $\bar{\Omega}'$ (measure ν supported by a neighbourhood of $\partial\Omega'$). If μ_n, μ correspond to

$$\hat{R}_\varphi^{e_n}, \hat{R}_\varphi^{\cup e_n} (e_n \subset \Omega'), \gamma(\cup e_n) = \int W \, d\mu = \int \hat{R}_\varphi^{\cup e_n} d\nu \leq \sum \int \hat{R}_\varphi^{e_n} d\nu$$

$= \sum \int W \, d\mu_n = \sum \gamma(e_n)$ and by a similar argument $\gamma(e_n) \to \gamma(\cup e_n)$.

If $\bar{e} \subset \Omega'$ and $\omega \supset e$, $\bar{\omega} \subset \Omega'$, we know that $R_\varphi^e = \inf R_\varphi^\omega$, hence $\hat{R}_\varphi^e = \inf \widehat{R_\varphi^{\omega_n}}$, for a suitable decreasing $\{\omega_n\}$ and by a $d\nu$-integration $\gamma(e) = \lim_n \gamma(\omega_n)$.

$\gamma(e)$ is therefore in Ω', a weight with the wanted properties and is even a general capacity.

If φ is superharmonic, $\gamma(K)$ satisfies, like $\hat{R}_\varphi^K$, the strong sub-additivity and is therefore a Choquet capacity. Thanks to the previous properties, we see that the inner, then the outer capacity of ω (resp. e) is $\gamma(\omega)$ (resp. $\gamma(e)$).

<u>Exercise:</u> For any $e \subset \bar{e} \subset \Omega$, there exists $\omega_n \downarrow$ open, $\supset e$ such that $\gamma(\omega_n) \to \gamma(e)$ and $\hat{R}_\varphi^{\omega_n} \to \hat{R}_\varphi^e$ (φ any finite continuous positive function). Same with any e, φ superharmonic and the outer capacity.

d) The characterisations of a polar set are easy consequences of the <u>criterion</u> $\hat{R}_\varphi^e \equiv 0$ $(\varphi > 0)$.

e) It remains to prove the Choquet property. Instead of the original proof of Choquet [6], we shall give another one which is an immediate consequence of the following result, important in itself.

5. Lemma VIII, 13. Given $e \subset \Omega$, $x_0 \in \Omega$, there exists a potential V, finite at x_0, such that at every $x \in \bar{e} \setminus \tilde{e}$, $V(y) \underset{y \in e,\, y \to x}{\longrightarrow} +\infty$.

Proof. We shall use the potential U of theorem VII, 6 (measure μ). Suppose first e is a base, the restrictions $\mu_1 = \mu|e$, $\mu_2 = \mu|\complement e$, have potentials u_1, u_2 finite continuous; $b^e_{\mu_1} = \mu_1$, $b^e_{\mu_2}(\complement e) = 0$. We introduce ω_n open $\supset \complement e$, such that $b^e_{\mu_2}(\omega_n) \to 0$ and $\nu_n = b^e_{\mu_2}|\omega_n$. We may extract a ν_{n_p} of potential v_p such that $\sum v_p(x_0) < +\infty$. $\sum v_p$ solves the question: for, $x \in \bar{e}$, $x \in \complement e$, $\liminf\limits_{\substack{y \in e \\ y \to x}} v_p(y) - v_p(x)$ is equal to

$$\liminf_{\substack{y \to x \\ y \in e}} \mathcal{B}^e_U(y) - \mathcal{B}^e_U(x) = U(x) - \mathcal{B}^e_U(x) > 0.$$

Hence the wanted property for $\sum v_p$. For any e, $e \subset \tilde{e} = B_e + $ a polar set α; the sum of the previous function for B_e and of a potential, infinite on $\alpha - \{x_0\}$, finite at x_0 solves the question. Because $\bar{e} = \bar{\tilde{e}} = \bar{B}_e \cup \bar{\alpha}$ and $\bar{e} \setminus B_e = (\bar{B}_e \setminus B_e) \cup (\bar{\alpha} \setminus B_e)$; if $x_0 \in e$, x of $\bar{e} \setminus \tilde{e}$ is $\neq x_0$.

Remark. Similarly there exists a potential W such that at every $x \in \bar{e} \setminus B_e$, $W(y) \underset{y \in e,\, y \to x}{\longrightarrow} +\infty$. Same proof by taking $x_0 \notin e$ (If $e = \Omega$, $\bar{e} \setminus B_e = \emptyset$).

Lemma VIII, 14. If ψ is ≥ 0, locally bounded and ω a variable open set containing $\bar{e} \setminus \tilde{e}$, $\inf\limits_{\omega} R^{e \cap \omega}_{\psi}(x_0) = 0$, $\forall x_0$.

Proof. For every $x \in \bar{e} \setminus \tilde{e}$ and every λ, there exists an open neighbourhood σ of x such that the previous V of the lemma satisfies $\psi < \lambda V$ on $\sigma \cap e$. If δ is the union of these σ, we get $\psi < \lambda V$ on $\delta \cap e$ and $R^{e \cap \delta}_{\psi}(x_0) < \lambda V(x_0)$. Hence $\inf\limits_{\omega} R^{e \cap \omega}_{\psi}(x_0) = 0$ (ω open $\supset \bar{e} \setminus \tilde{e}$) and that is true for every x_0.

Same result and proof (or consequence) for $\omega \supset \bar{e} \setminus B_e$ with $x_0 \notin e$.

The proof of the Choquet property is now easy. We will give it even in larger conditions, as follows.

Theorem VIII, 15. If ψ is ≥ 0 locally bounded, $R^e_{\psi}(x_0)$ (any fixed x_0), $\int R^e_{\psi}\, dm(x)$ ($m \geq 0$ does not charge polar sets and $\int G_x\, dm \not\equiv +\infty$) and $\mu(\Omega)$ where μ is the measure corresponding to $\hat{R}^e_{\psi}$ are weights of Choquet type, the last one in any $\Omega' \subset \bar{\Omega}' \subset \Omega$.

Proof. The increasing property of these set functions is a consequence of the increasing property of R^e_ψ. The first weight is c.t according to the previous lemma, where ω may be replaced by $\omega' = \omega \cup \complement \bar{e}$. Moreover, there exists ω'_n decreasing, containing $\complement \tilde{e}$, such that, $\forall x$

$$\inf \widehat{R}_\psi^{e \cap \omega'_n}(x) = 0, \text{ hence } R_\psi^{e \cap \omega'_n} \to 0 \text{ q.e.}$$

By using $\Omega_p \uparrow$, $\bar{\Omega}_p \subset \Omega$, $\bigcup \Omega_p = \Omega$, we get $\int R_\psi^{e \cap \omega'_n \cap \Omega_p} dm \leq \lambda \int G_x(y) dm(y)$ for a suitable λ, $\forall n$, fixed p, therefore $\int R_\psi^{e \cap \omega'_n \cap \Omega_p} dm \to 0$. We choose $n = n_p$ such that this integral is $< \varepsilon 2^{-p}$. Then $\int R_\psi^{e \cap (\bigcup \omega'_{n_p} \cap \Omega_p)} dm < \varepsilon$. The question is solved for the second weight. Finally, $\widehat{R}_\psi^{e \cap \omega'_n \cap \Omega'}$ is the potential of a measure μ_n (on $\bar{\Omega}'$). By using the same W of (c) (proof of the VIII, 12), $\int \widehat{R}^{e \cap \omega'_n \cap \Omega'} d\nu = \int W d\mu_n = \mu_n(\Omega) \to 0$. Hence, the Choquet property in Ω' for the third weight.

Remark. The interpretation of polar sets as sets of weight $A_m(e)$ zero for a suitable m leads to the result.

Any fine closed set is upto an additive polar set, a F_σ -set (see Prop. IV, 5) as well as a G_δ -set (VII, § 3).

6. Some indications on various complements.

a) The role of fine topology is important in the study of weights of increasing and chiefly decreasing sets. For instance, the capacities $A_m(e)$, $\gamma(e)$ of Th. VIII, 12, for decreasing fine closed sets e_n, contained in a compact K, tend to the corresponding capacities of $\bigcap e_n$, as a consequence of $\widehat{\inf \widehat{R}_\varphi^{e_n}} = \widehat{R}_\varphi^{\bigcap e_n}$ and that may be extended to directed families $\{e_i\}$ (see Brelot [30]).

A deeper study of the weights $R_1^e(x_0)$, $\int \widehat{R}_\psi^e dm$ (for any $\psi \leq V$ superharmonic satisfying $\int V dm < \infty$, and m not charging polar sets, for decreasing e, is developed in Brelot [27, 30, 32, 33], chiefly when ψ is superharmonic ≥ 0. These weights are of Choquet type.

b) Statistical thinness. We have met (Th. VIII, 12 and 15) capacities which are weights of Choquet type. As an easy consequence

<u>Theorem VIII, 16</u> (Brelot [27]). <u>The following properties are equivalent</u>

a) <u>On</u> Ω, e <u>is thin</u> q.e. <u>on the set</u> α (<u>statistical thinness on</u> α)

b) $\inf\limits_{\omega} \widehat{R_1^{e\cap\omega}} = 0$ (<u>at one point or at every point, or equivalently with</u> R_1)
i.e. <u>the family</u> $\{e\cap\omega\}$ <u>is 1-vanishing</u> (ω, a variable neighbourhood of α).

In fact, thanks to the Choquet property, $\{e\cap\delta\}$ is 1-vanishing for the neighbourhoods δ of $\complement\tilde{e}$ i.e. of the set of points of $\complement e$ where e is thin; and starting from the statistical property (a), it is the same for the neighbourhoods of the following polar sets.

1) part of α where e is not thin, 2) part of e where e is thin. Hence (b).

The converse is easy: We start from (b); if α_0 is the part of α where e is unthin and ω a variable neighbourhood of α, $e\cap\omega$ is unthin on α_0 and, because

$$R_1^{\tilde{\gamma}} = R_1^{\gamma} \quad (\tilde{\gamma}, \text{ fine closure of } \gamma), \text{ we get}$$

$$R_1^{(e\cap\omega)\cup\alpha_0} = R_1^{e\cap\omega}, \quad R_1^{\alpha_0} \leq R_1^{e\cap\omega}$$

$$R_1^{\alpha_0} \leq \inf R_1^{e\cap\omega}, \quad \hat{R}_1^{\alpha_0} \leq \inf \widehat{R_1^{e\cap\omega}} .$$

Hence α_0 is polar.

<u>Remark</u>. Equivalent conditions are:

a) 1-vanishing of the fine neighbourhoods of α.

b) Existence of a superharmonic function $V > 0$ whose restriction to e tends to $+\infty$ at the points of $\alpha\cap\bar{e}$ (same result with fine limit at points of $\alpha\cap\tilde{e}$). See lemma VIII, 13.

c) <u>Some definitions</u> and applications (see Brelot [27, 30, 33]).

A) For W superharmonic > 0 (with associated measure μ_W), a W-polar set on Ω is a polar set e such that outer $\mu_W(e) = 0$. Equivalently the neighbourhoods of e form a W-vanishing family.

<u>application</u> Equivalences.

1) e is thin on α except on a W-polar set (W-statistical thinness on α).

2) For the neighbourhoods ω of α on Ω (as well fine neighbourhoods) $\{e\cap\omega\}$ is W-vanishing.

3) Existence of U superharmonic > 0 such that $\frac{U(y)}{W(y)}$ (taken as $+\infty$ when $U = W = +\infty$) tends to $+\infty$. ($y \in e$, $y \to x \in \bar{e}$).

Same property with fine convergence and closure.

B) A superharmonic function $W > 0$ on Ω is <u>semi-bounded</u> if the family of $e_\lambda = \{x \mid W > \lambda\}$ is W-vanishing. When W is a potential, this is equivalent to the condition that the set $\{W = +\infty\}$ is W-polar.

<u>application</u>. Extension of the first result of previous (a).

If $\{\varphi_i\}$ is a directed decreasing set of fine upper semi-continuous functions ≥ 0 on Ω, majorized by a fixed semi-bounded potential V,

$\widehat{\inf \hat{R}^V_{\varphi_i}} = \hat{R}_{\inf \varphi_i}$.

Chapter IX

Further study of classical thinness(33). Some applications

1. We have shown the interest of the notion of thinness for general theories like balayage, which may be extended to axiomatic developments. Therefore, it is natural to deeper the study in the classical case which will offer more details and examples, useful in classical applications. We will consider chiefly the euclidean space and it is easy to make extensions to the neighbourhood of the point at infinity by similar arguments or by using inversion and Kelvin transforms. Then, the general Green space may be considered, directly or otherwise.

Recall that in a Green space, thinness implies strong thinness and hyper-thinness.

Remarks on the preservation of thinness or unthinness by mapping.

Consider in $\mathbb{R}^n$ a bounded borel mapping $x \rightsquigarrow y = F(x)$ (bounded for x in any ball). A measure ν may be associated to any given Radon measure μ by the condition $\int \theta(y)\, d\nu(y) = \int \theta(F(x))\, d\mu(x)$, for any θ finite continuous with compact support. ν is called the image of μ by F and the formula holds for any $d\nu$-summable θ.

Proposition IX, 1. When F is bijective and preserves the distances $|x_1 - x_2|$ upto a coefficient $\lambda(x_1,x_2)$ $(0 < \alpha \leq \lambda(x_1,x_2) \leq \beta$ finite) (x_1, x_2 in a neighbourhood of x_0), thinness at $x_0 \Longleftrightarrow$ thinness of the image at $F(x_0) = y_0$.

Proof: $\int h(|y - y_1|)\, d\nu(y) = \int h(\lambda(x, x_1)\, |x - x_1|)\, d\mu(x)$. If the potential of μ is finite at x_0 and tends to $+\infty$ on e ($x \to x_0$, $x \neq x_0$), the potential of ν is finite at y_0 and tends to $+\infty$ on $F(e)$ ($y \to y_0$, $y \neq y_0$) and conversely.

Proposition IX, 2. If F preserves the distances to an invariant point x_0, but diminishes the distances in general, potentials are increased therefore a thin set at x_0 becomes a thin set at $y_0 = x_0$.

---

(33) See chiefly Brelot [4], [5], [8], Cartan [2] with bibliography.

Example. In $\mathbb{R}^2$, the mapping $x \to y$ where y is on a half line issued from x_0 and $|y - y_0| = |x - x_0|$.

These remarks will allow to build more examples from elementary ones. We will indicate now.

2. A geometrical property of thinness:

Theorem IX, 3. If $e \subset \mathbb{R}^n$ is thin at x_0 and if $\sigma^r_{x_0}$ denotes the unitary positive spherical measure on the sphere $\partial B^r_{x_0}$ (uniform, i.e. propositional to the area), then $\sigma^{r*}_{x_0}(e \cap \partial B^r_{x_0}) \to 0$ as $r \to 0$ (outer measure).

Proof. We may suppose $x_0 \notin e$, e open and the existence of a superharmonic function $u > 0$ in a neighbourhood of x_0 such that $u(x) \to +\infty$, $x \in e$, $x \to x_0$, $u(x_0)$ finite. Now,

$$u(x_0) \geqslant \text{mean of } u \text{ on } \partial B^r_{x_0} \geqslant \lambda(r)\, \sigma^{r^*}_{x_0}(\partial B^r_{x_0} \cap e)$$

where $\lambda(r) = \inf u$ on $e \cap \partial B^r_{x_0}$. As $\lambda(r) \to +\infty$ as $r \to 0, \sigma^{r^*}_{x_0}(\partial B^r_{x_0} \cap e) \to 0$.

Exercise: Similar proof with u satisfying only $u(x_0) < \liminf\limits_{\substack{x \to x_0 \\ x \in e}} u(x)$.

Extension. In a Green space, if $\{\omega_i\}$ is a directed decreasing family of domains containing x_0, with intersection $\{x_0\}$ and if $d\rho^{\omega_i}_{x_0}$ is the harmonic measure at x_0, then $\inf\limits_i \rho^{*\omega_i}_{x_0}(e \cap \partial\omega_i) = 0$.

Similar proof.

At a point at infinity, this implies on the local image on $\overline{\mathbb{R}}^n$ a property similar to the theorem, with a sphere of fixed centre and radius $\to \infty$.

Remark. Any bounded function in a neighbourhood of x_0, which has a fine limit λ at x_0, has a mean value on $\partial B^r_{x_0}$ tending to λ $(r \to 0)$.

Example of thinness. Any set of $\mathbb{R}^n$ containing a cone of revolution (with interior $\neq \emptyset$) is unthin at the vertex. (actually indicated by H. Poincaré about the Dirichlet problem).

3. Criteria of thinness and unthinness.

Theorem IX, 4. A closed set $e \subset \mathbb{R}^n$ is unthin at $x_0 \in \partial e$ iff there exists

<u>on</u> $\sigma \setminus e$ (<u>suitable open neighbourhood</u> σ <u>of</u> x_0), <u>a superharmonic positive</u> (>0) <u>function</u> u <u>tending to zero at</u> x_0.

<u>Proof</u>. We will use a superharmonic finite continuous function $V > 0$ in a ball B_{x_0}, which is not harmonic in any open subset (for instance, $k - |x - x_0|$ or $k - |x - x_0|^2$ for a suitable constant k or the potential of the Lebesgue measure on B_{x_0}). Consider the reduced functions relative to B_{x_0} and $e' = e \cap B_{x_0}$. If e is unthin at x_0, $\hat{R}_V^{e'}(x_0) = V(x_0)$, therefore, $\hat{R}_V^{e'}$ is continuous at x_0 and $V - \hat{R}_V^{e'}$ which is $\gg 0$ cannot be zero in any component of $\complement e'$ in B_{x_0}, therefore is > 0. It satisfies the wanted conditions.

Conversely, suppose v is superharmonic on $B_{x_0} \setminus e$ for a suitable ball, with $v > 0$, $v \to 0$ at x_0. Let us see that it is incompatible with the thinness of e, i.e. with the existence of a subharmonic function u in B_{x_0} satisfying $u(x_0) = 1$, $u(x) \leq -1$ on $e \cap B_{x_0}^r$ (suitable r).

We observe that $u \leq v$ on an open neighbourhood δ of $\partial B_{x_0}^r \cap e$ and $u \leq \lambda v$ on $\partial B_{x_0}^r \cap \complement \delta$ ($\lambda > 1$ large enough). Therefore $u - \lambda v - \varepsilon G_{x_0}^{B_{x_0}}$ ($\varepsilon > 0$) on $B_{x_0}^r \setminus e$ has a lim sup ≤ 0 at every boundary point. Hence, it is ≤ 0 and $u \leq \lambda v$ on $B_{x_0}^r \setminus e$. That is incompatible with $u(x_0) = 1 = \limsup_{\substack{x \neq x_0 \\ x \to x_0}} u(x)$.

<u>Extension to the thinness in a Green space</u>.

Easy adaptation.

<u>Applications</u>. 1) <u>Proposition</u> IX, 5. <u>A boundary point</u> $\neq \mathcal{A}$ <u>of an open set</u> ω <u>in a Green space</u> Ω <u>is irregular iff</u> $\complement_\Omega \omega$ <u>is thin</u>. (contained in theor VII, 13)

<u>Proof</u>. Because, there are identical local criteria (Chap. VI, § 6; Th. IX, 4).

Hence another proof that the set of these points is polar.

2) <u>Some Examples</u>. In $\mathbb{R}^n$, a <u>hyperplane</u> H ((n-1) dim) is unthin at its points, because the distance to H is harmonic > 0 outside H. But a linear manifold of dimension $\leq n-2$ is locally polar (because the Newtonian or the log potential of the Lebesgue measure of an open part α is infinite on α), therefore is thin everywhere.

Half an hyperplane is unthin even on the edge (consider the union with the other half hyperplane). In $\mathbb{R}^2$, a segment is unthin at its points.

3) <u>Proposition IX, 6.</u> <u>In</u> $\mathbb{R}^2$ <u>for any thin set</u> e <u>at</u> x_0, <u>there exist arbitrary small circles</u> $|x - x_0| = r$, which do not intersect e (improved Lebesgue-Beurling property).

<u>Proof</u>. Because the mapping of § 1 for a half-line issued from x_0 gives a thin set which cannot contain any segment one of whose ends is x_0.

Therefore, in $\mathbb{R}^2$, for a Greenian relatively compact finitely connected open set, all non-isolated boundary points are regular.

In $\mathbb{R}^n$ $(n \geqslant 3)$, there is nothing similar. Let us recall the famous example of the <u>Lebesgue spine</u> in $\mathbb{R}^3$ (actually given as an example of an irregular boundary point).

We consider on the segment $[0, 1]$ of the x_1-axis, a measure of linear density equal to x_1. The Newtonian potential V is 1 at the origin 0, $+\infty$ on the segment. The set $\{x : V > K > 1\}$ (of revolution around the x_1-axis) is thin at 0. Elementary calculations show that the meridian limit - curve has an equation $z = \pm\, a^{-\frac{\lambda(x_1)}{x_1}}$ where $\lambda(x_1) \to \frac{K}{2}$.

4. <u>Theorem IX, 7.</u> <u>At</u> $x_0 \in \Omega$, (Greenian space), <u>the thinness of</u> e <u>is equivalent to</u>:

(α) : $\mathcal{B}^e_{G_{x_0}} \neq G_{x_0}$, i.e. $b^e_{\varepsilon_{x_0}} \neq \varepsilon_{x_0}$ or also to $b^e_{\varepsilon_{x_0}}\{x_0\} = 0$(34).

<u>Other forms in</u> $\mathbb{R}^n$ <u>of this criterion are</u>

a) <u>There exists in the neighbourhood of</u> x_0, <u>a superharmonic function</u> v <u>such</u>

---

(34) Improving a notion of De La Vallée Poussin, Cartan [2] (in $\mathbb{R}^n$, $n \geqslant 3$, or in the unit disc of $\mathbb{R}^2$) takes the property $b^e_{\varepsilon_{x_0}} \neq \varepsilon_{x_0}$ as the definition of the "enterior regularity" of x_0 for e (therefore, equivalent to the thinness of e at x_0). With the "second type balayage" (see footnote 24), he gets similary an "interior regularity" which is equivalent to the fact that any closed set of $e \cup \{x_0\}$ is thin at x_0 (interior thinness). This second notion is less useful and we set it aside here, as well as the second type balayage. We reserve here the word regularity for a previous and classical notion in the Dirichlet problem, in order to avoid any confusion.

<u>that</u>

$$\liminf_{\substack{x\in e\\ x\neq x_0, x\to x_0}} \frac{v(x)}{h_{x_0}} > \liminf_{\substack{x\neq x_0\\ x\to x_0}} \frac{v(x)}{h_{x_0}} \quad \text{(recall that } h_{x_0} \text{ means } h(|x - x_0|)).$$

b) <u>There exists in such a neighbourhood a measure</u> $\mu \geqslant 0$, <u>which does not charge</u> $\{x_0\}$ <u>and whose potential</u> (h <u>or</u> G-<u>potential for any domain</u>) <u>satisfies</u>

$$\liminf_{\substack{x\in e\\ x\neq x_0, x\to x_0}} \frac{v(x)}{h(|x-x_0|)} > 0.$$

<u>Equivalently</u> $v \geqslant h$ or G_{x_0} <u>on</u> e <u>close to</u> x_0, <u>or</u> only q.e. and we may realize $\mu(\complement B_e) = 0$.

c) <u>There exists in such a neighbourhood a measure</u> $\mu \geqslant 0$ <u>whose</u> G <u>or h-potential</u> v <u>satisfies</u> $\frac{v(x)}{h(|x-x_0|)} \to +\infty \ (x \in e\ ,\ x \neq x_0)$.

<u>Proof</u>. If e is thin, $b^e_{\varepsilon_{x_0}}$ which is supported by B_e does not charge $\{x_0\}$. If e is unthin (i.e. if $x_0 \in B_e$), $b^e_{\varepsilon_{x_0}} = \varepsilon_{x_0}$.

Let us first see the equivalence of (a), (b), (c).

We will use the <u>following remark</u>. If in a Green space Ω, v is a potential majorizing $G^{\Omega}_{x_0}$ (x_0 polar), then $\mu_v(\{x_0\}) \neq 0$. If not, v would be in $\Omega\setminus\{x_0\}$, a $G^{\Omega\setminus\{x_0\}}$ potential with a harmonic minorant $G_{x_0} \neq 0$.

Starting from (a), we deduce the same inequality for the local potential part of v, then for the potential V of the restriction of the measure to $\complement\{x_0\}$. Then the new second member must be zero (if not, λV with a suitable λ would majorize some $G^{B_{x_0}}_{x_0}$ close to x_0, hence in B_{x_0} and the previous remark gives a contradiction). We thus got (b).

Now (b) $\Rightarrow$ (a), because $\frac{v}{h(|x-x_0|)}$ or $\frac{v}{G^{B_{x_0}}_{x_0}}$ has a lim inf zero (same argument). As (c) $\Rightarrow$ (b), we have only to prove that (b) $\Rightarrow$ (c).

If the v of (b) has a corresponding measure μ, consider the restriction μ_n to $B^{\gamma_n}_{x_0}$ and a subsequence n_p such that $\sum \|\mu_{n_p}\|$ is finite. Then the measure $\sum \mu_{n_p}$ answers the question.

Let us prove finally that the thinness of e at x_0 in a ball Ω is equivalent to (b). Supposing the thinness, $b^e_{\varepsilon_{x_0}}$ does not charge $\{x_0\}$ and its G^{Ω}-potential, is equal to G_{x_0} on e, except on a polar set e. Let us introduce on Ω a superharmonic function $w > 0$, $+\infty$ on $e \sim \{x_0\}$, finite at x_0. Then,

$$\liminf_{\substack{x \in e \\ x \neq x_0, x \to x_0}} \frac{\mathcal{B}^e_{G_{x_0}} + w}{G_{x_0}} > 0. \quad \text{Hence (b).}$$

Starting from (b), denote by λ, the lim inf and for $0 < \lambda' < \lambda$, $\frac{v}{\lambda'}$ majorizes G_{x_0} on $e \cap \sigma$, for a suitable neighbourhood σ of x_0. Hence $\frac{v}{\lambda'} \geqslant \mathcal{B}^{e\cap\sigma}_{G_{x_0}}$. $\mathcal{B}^{e\cap\sigma}_{G_{x_0}}$ cannot be equal to G_{x_0} (by using the preliminary remark). Hence, the thinness of $e \cap \sigma$ and therefore of e.

<u>Remark</u>: By using only the additive property of $v \to \mathcal{B}^e_v$ (v superharmonic $\geqslant 0$ and the invariance $\mathcal{B}^e_{\mathcal{B}^e_v} = \mathcal{B}^e_v$, we may prove immediately that $\mathcal{B}^e_{G_{x_0}} \neq G_{x_0}$ implies $b^e_{\varepsilon_{x_0}}(\{x_0\}) = 0$ (in any Green space). Therefore, the previous argument shows that (a), (b), (c) are equivalent to $\mathcal{B}^e_{G_{x_0}} \neq G_{x_0}$, independantly of the result (Th. VIII, 3) of balayage theory.

<u>Exercise</u>. Prove the equivalence of (a), (b), (c) with thinness by using the following lemma (useful, later too). So is more directly proved, the criterion (α) of thinness in $\mathbb{R}^n$ (See Brelot [5]).

<u>Lemma IX, 8.</u> In $\mathbb{R}^n$, consider a measure $\mu > 0$, not charging the origin 0, its potential (log or newtonian) v and a number $s > 1$. The restriction of μ on the set

$$E_n = \{x \mid h(|x|) \leq s^{n-1}\} \cup \{x \mid h(|x|) \geqslant s^{n+2}\}$$

has a potential V which is on the set: $I_n = \{x \mid s^n \leq h(|x|) \leq s^{n+1}\}$,

(A) majorized by $h(|x|).\varepsilon_n$ ($\varepsilon_n \to 0$) and also

(B) when $v(0)$ is finite, by $K.v(0)$, where K is independant of μ and n.

(Wiener-De La Vallée Poussin).

Elementary calculations show that $\frac{h(|x-y|)}{h(|x|)}$ and $\frac{h(|x-y|)}{h(|x-y_1|)}$ are bounded for variable $x \in E_n$ and $y \in I_n$ and n. The second wanted result is then obvious. The first one is a consequence of the known property(35) that the mean value of v on ∂B_0^r has an expression $h(r).\varepsilon(r)$ ($\varepsilon(r) \to 0$, for $r \to 0$).

5. Theorem IX, 9 (Bouligand). If Ω is any Green space, a closed set e is unthin at $x_0 \in e$, iff a) the solution of the Dirichlet problem for $\Omega \setminus e$, with boundary values $G_{x_0}^{\Omega}$ and 0 at the Alexandroff point of Ω is equal to $G_{x_0}^{\Omega}$, or when x_0 is polar iff b) there exists on $\sigma \setminus e$, for an open neighbourhood of x_0, a superharmonic function u such that $u/G_{x_0}^{\Omega} \to +\infty$, $x \in \complement e$, $x \to x_0$. (equivalently, if x_0 is not a point at infinity, to the condition on the local image $\frac{u}{h(|x-x_0|)} \to +\infty$, $x \in \complement e$, $x \to x_0$).

Proof. (a) is in our case of e closed, the same as criterion (α) of th IX, 7.

Suppose e unthin at x_0, polar. We will even see that there exists in $\Omega \setminus e$ a harmonic function u, > 0, such that $\frac{u}{G_{x_0}^{\Omega}} \to +\infty$ at x_0. For every domain $\sigma \subset \Omega$, $\sigma \ni x_0$, $(\hat{R}_{G_{x_0}^{\sigma}}^{e \cap \sigma})_{\sigma} = G_{x_0}^{\sigma}$ on σ. Now, for the function f_σ equal to $G_{x_0}^{\Omega}$ on σ, o elsewhere and particularly at the Alexandroff point of Ω, $H_{f_\sigma}^{\Omega \setminus e} \geqslant \hat{R}_{G_{x_0}^{\sigma}}^{e \cap \sigma}$ on $\sigma \setminus e$. Starting from a sequence σ_n (decreasing with $\bigcap \sigma_n = \{x_0\}$), we choose a subsequence σ_p^1 for which $\sum_1 H_{f_{\sigma_p^1}}^{\Omega \setminus e}(x_1)$ is finite(36) (x_1 in the component ω_1 among all components ω_i of $\Omega \setminus e$), then a subsequence σ_p^2 of the previous one such that $\sum_1 H_{f_{\sigma_p^2}}^{\Omega \setminus e}(x_2)$ is finite ($x_2 \in \omega_2$)..... By the diagonal process, we get a subsequence σ_p' such that

---

(35) Instead of some ordinary proof, one may use the graph of the mean on B_0^r as a concave function of $t = h(r)$, the interpretation of the right half-tangent as the corresponding mass supported by B_0^r (upto a factor). This property implies in $\mathbb{R}^n$, the preliminary remark used in the proof of theorem IX,7.

(36) As $f_{\sigma_p^1} \to 0$ on the boundary except on x_0 (polar), the general term of the series may be made $< 2^{-p}$.

$u = \sum H_{f_{\sigma'_p}}^{\Omega\setminus e}$ is harmonic in $\Omega\setminus e$; and now, $u \geqslant N\, G_{x_0}^{\sigma'_N}$ on $\sigma'_N - e$.

$\geqslant \frac{N}{2} G_{x_0}^{\Omega}$ on a suitable neighbourhood of x_0[37] (outside of e), which is giving the wanted behaviour of u at x_0.

Conversely, suppose the existence of u in condition (b). Then $\varepsilon u - G_{x_0}^{\Omega} + H_{G_{x_0}^{\Omega}}^{\sigma\setminus e}$ ($\varepsilon > 0$) has a lim inf $\geqslant 0$ at every regular boundary point of $\sigma\setminus e$ ($\bar{\sigma}\subset\Omega$). Hence $G_{x_0}^{\Omega} - H_{G_{x_0}^{\Omega}}^{\sigma\setminus e} \leqslant \varepsilon u$ and hence the condition (a).

Exercise. Among various other forms of the criterion, prove the following one: for one or any relatively compact open neighbourhood σ, any non-negative harmonic function in $\sigma\setminus e$, tending to zero q.e at the boundary and majorized by $G_{x_0}^{\Omega}$ + constant, must be zero.

6. Theorem IX, 10. (Famous Wiener criterion)[38]. If Ω is any Greenian domain of $\mathbb{R}^n$, (for instance $\mathbb{R}^n$ for $n \geqslant 3$) and if $x_0 \in \Omega$, $\gamma(\alpha)$ the corresponding classical outer capacity, i.e. when α is relatively compact, the measure of Ω, for the measure corresponding to $\mathcal{B}_1^{\alpha}$, and

$$I_n, \text{ the set } \left\{x \,\middle|\, s^n \leqslant h(|x - x_0|) \leqslant s^{n+1}\right\} \quad (s > 1),$$

then the thinness of any e at x_0 is equivalent to the condition $\sum \gamma(I_n \cap e) s^n$ finite or equivalently to $\sum \mathcal{B}_1^{I_n \cap e}(x_0)$ finite.

These last conditons are equivalent, because the measure corresponding to $\mathcal{B}_1^{I_n \cap e}$ is supported by I_n and its potential at x_0 is between $\gamma(I_n \cap e)\, s^n$ and $\gamma(I_n \cap e)\, s^{n+1}$.

Proof. Now this condition implies that $\mathcal{B}_1^{e \cap B_{x_0}^r}(x_0) \to 0$ as $r \to 0$, i.e. that e is strongly thin. The converse is more difficult. We start from a G^{Ω}-potential v, finite at x_0, but tending to $+\infty$ on e at x_0; if $E = \bigcup_p I_{2p} \cap e$, $\mathcal{B}_v^E$

---

(37) Because $G_{x_0}^{\sigma'_N}/G_{x_0}^{\Omega}$ tends to 1 at x_0.

(38) Originally (Wiener [2]) given as a criterion of irregularity of a boundary point of an open set. The proof was complicated because of the lacking of advanced balayage theory. The useful part of lemma IX,8 was already introduced.

is finite at x_0, tends to $+\infty$ on E except on a polar set; its measure μ is supported by $\bar{E}$ and the restriction $\mu \mid \complement I_{2p}$ has a h or G-potential bounded on $I_{2p} \cap e$ (lemma IX, 8, B); then $\mu \mid I_{2p}$ for $p \geqslant p_0$ large encough has a potential $V_p \geqslant 1$ on $I_{2p} \cap e$, except on a polar set. Therefore, $V_p \geqslant \mathcal{B}_1^{I_{2p} \cap e}$ for $p \geqslant p_0$ and $\sum_{p_0}^{\infty} \mathcal{B}_1^{I_{2p} \cap e}(x_0) \leq$ G-potential of μ at $x_0 = \mathcal{B}_v^E(x_0)$ finite. Similar argument for $\mathcal{B}_1^{I_{2p+1} \cap e}(x_0)$ and hence the wanted convergence.

Remark 1. Same result with the sets $\{s^n < h(|x - x_0|) \leq s^{n+1}\}$ or $\{s^n \leq h(|x - x_0|) < s^{n+1}\}$ instead of I_n. Immediate adaptation.

Remark 2. Integral forms of the criterion have been given; first for closed sets (Kellogg-Vasilesco-Frostman) as a consequence or by direct arguments. If $\delta(z)$ is the set $e \cap \{x \mid h(|x - x_0|) \geqslant z\}$, it is the condition $\int \gamma(d(z))dz$ finite. But that seems less useful.

However, by using it, one may easily deeper the study of Th IX, 3 (see Deny [1]) and precise the richness of the circles $|x_0 - x| = r$ which do not intersect the set thin at x_0 (application 3 of Th. IX, 4) (see Brelot [4]).

Exercise. Prove that in $\mathbb{R}^n$, if a function f has a fine limit 1 at x_0, it tends to 1 along all half-lines issued from x_0 except those of a cone intersecting the unit sphere (centre x_0) according to a locally polar set. (result of Deny [1]).

Complements. a) Another form of the criterion is the convergence of the series of general term $\mathcal{B}_{h(|x-x_0|)}^{e \cap I_n}(y_0)$, $y_0 \neq x_0$.

b) Other forms are obtained by considering instead of I_n, the set

$$J_n = \{t^{n+1} \leq |x-x_0| \leq t^n\} \quad 0 < t < 1$$

(equivalently latter with $<$ instead of any $\leq$). We may now replace the previous reduced functions relative to $e \cap I_n$ by the same relative to $e \cap J_n$. In the case of $\mathbb{R}^n (n \geqslant 3)$, there is nothing new. In the case of $\mathbb{R}^2$, it is the same as the convergence of $\{n\,\gamma(J_n \cap e)\}$ (Tsuji). Proofs may be given by adaptation of the previous one.

<u>Remark</u>. The condition $R_1^{e \cap J_n}(x_0) \to 0$ or $R^{e \cap J_n}_{h(|x-x_0|)}(y_0) \to 0$ may be taken as a definition of <u>semi-thinness</u> (independant of t). In $\mathbb{R}^2$, that means $n\,\gamma(J_n \cap e) \to 0$.

c) In $\mathbb{R}^n (n \geqslant 3)$, thinness at the point at infinity may give interesting forms of criterion. For example, it is equivalent to the fact that the outer classical capacity is finite (see Brelot [6, 8, 10]).

7. <u>Some applications of thinness and fine limits</u>. (See essentially Brelot [5 to 10])

<u>Lemma IX, 11</u>. For a lower bounded superharmonic function on an open set $\omega \subset \mathbb{R}^n$, denote for $x \in \partial\omega$, $\mathcal{L}_x u = \liminf\limits_{\substack{y \in \omega \\ y \to x}} u(y)$. If $x_0 \in \partial\omega$ is not isolated on $\partial\omega$ and if $\mathcal{L}_{x_0} u < \liminf\limits_{\substack{x \in \partial\omega \\ x \neq x_0, x \to x_0}} \mathcal{L}_x u$, then $\complement\omega$ is thin at x_0 and u has at x_0, a fine limit equal to $\mathcal{L}_{x_0} u$.

<u>Proof</u>. Let K be strictly between both limits. We consider inf (u, K) on ω and define a continuation by the value K on $\complement\omega \setminus \{x_0\}$. We get a superharmonic function which becomes superharmonic on the neighbourhood of x_0 by giving at x_0 the value $\mathcal{L}_{x_0} u$. The set where this function U is equal to K, is thin and contains $\complement\omega \setminus \{x_0\}$; and U is fine continuous at x_0. As $u = U$ on a fine neighbourhood of x_0, less x_0, we get the wanted result.

<u>Theorem IX, 12</u>. <u>If</u> u <u>is superharmonic on</u> ω (<u>open set of a Green space</u>), <u>lower bounded close to</u> $x_0 \in \partial\omega$, <u>with</u> $\complement\omega$ <u>thin at</u> x_0 (i.e. x_0 <u>irregular for</u> ω) <u>then</u> u <u>has a fine limit at</u> x_0.

<u>Proof</u>. We come back to the case of ω bounded in $\mathbb{R}^n$ and u lower bounded. Result obvious if x_0 is isolated on $\partial\omega$. If not, we introduce v superharmonic in a neighbourhood of x_0 with $v(x_0)$ finite and $v(x) \to +\infty$ ($x \in \complement\omega$, $x \neq x_0$, $x \to x_0$). If $u + v$ tends to $+\infty$ on ω ($x \to x_0$), therefore also in the fine topology, as v has a finite fine limit, u has an infinite fine limit at x_0.

If $\mathcal{L}_{x_0}(u + v)$ is finite, let us note that $\mathcal{L}_x(u + v) \to +\infty$ ($x \neq x_0$, $x \in \partial\omega$, $x \to x_0$) and the above lemma shows that $u + v$ has a fine limit equal to $\mathcal{L}_{x_0}(u+v)$ and therefore u has a fine limit.

Remark. It is obviously sufficient u be lower bounded on a fine neighbourhood contained in ω, because it is contained in an open set where u is also lower bounded.

Lemma IX, 13. If u is superharmonic in a neighbourhood of $x_0 \in \mathbb{R}^n$, u/h_{x_0} has a finite fine limit at x_0 ($x \neq x_0$, $x \to x_0$) equal to $\liminf\limits_{\substack{x \neq x_0 \\ x \to x_0}} \frac{u}{h_{x_0}}$.

Proof. We have seen that this lim inf $\frac{u}{h}$ is zero when $\mu_u\{x_0\} = 0$, therefore always is finite (See Th IX, 4 proof). If K is greater than the lim inf, then the set where $\frac{u}{h_{x_0}} > K$ must be thin (Th. IX, 7 a) and that proves that $\frac{u}{h_{x_0}}$ has a fine limit equal to the lim inf.

Lemma IX, 14. If u is superharmonic on $\omega \subset \mathbb{R}^n$ and satisfies

$$-\infty < \lambda = \liminf_{\substack{x \in \omega \\ x \to x_0}} \frac{u}{h_{x_0}} < \liminf_{\substack{y \in \partial\omega,\ y \neq x_0 \\ y \to x_0}} \frac{\mathcal{L}_y u}{h_{x_0}(y)} \quad (x_0 \text{ not isolated on } \partial\omega). \text{ Then}$$

$\complement\omega$ is thin at x_0 and $\frac{u}{h_{x_0}}$ has a fine limit at x_0 equal to λ.

Proof. We come back to $u \geq 0$ by adding $a.h$ $(a > \lambda)$ and we may give a proof similar to lemma IX, 11. We introduce K strictly between both lim inf, then consider $\inf(u, K h_{x_0})$ continued by Kh_{x_0} outside $\{x_0\}$. This function may be continued at x_0 as a superharmonic function U in the neighbourhood of x_0; the set $\left\{x \neq x_0,\ \frac{U(x)}{h_{x_0}(x)} = K\right\}$ is thin at x_0 (Th. IX, 13, a) and $\frac{U}{h_{x_0}}$ has the fine limit λ (lemma IX, 13). Hence the wanted result.

Theorem IX, 15. If u is superharmonic on ω (open set of a Green space Ω) ≥ 0 close to $x_0 \in \partial\omega$, with $\complement\omega$ thin at x_0, then $\frac{u}{G^{\Omega}_{x_0}}$ has a finite fine limit at x_0.

Proof. We come back to a bounded ω in $\mathbb{R}^n$. Proof is similar to IX, 12, using th IX, 7,c and the previous lemma.

Note that $u \geq 0$ may be replaced by $\frac{u}{G^{\Omega}_{x_0}}$ lower bounded on a fine neighbourhood of x_0. Of course, $G^{\Omega}_{x_0}$ may be replaced by any $G^{\Omega'}_{x_0}$, Ω' domain $\ni x_0$

in th. IX, 15.

Consequence. The solution H_f^{ω} of a Dirichlet problem in the neighbourhood of an irregular point x_o, has therefore a fine limit (as well as $H_f^{\omega}/G_{x_o}^{\Omega}$; these limits are actually 0) at least for $f \geqslant 0$; that may be deepened and the fine limit may be expressed as $\int f\, d\nu$ where ν is $b_{\varepsilon_{x_o}}^{\complement\omega}$ (obvious when f is a potential on Ω).

Remark. The behaviour of a bounded harmonic function u on Ω in a neighbourhood of an irregular x_o may be deepened; $u(x_n)$ tends to the fine limit at x_o, when $x_n \in \Omega$, $x_n \to x_o$ and $G_{y_o}^{\Omega}(x_n) \to$ fine limit of $G_{y_o}^{\Omega}$ at x_o ($y_o \in \Omega$) i.e lim sup $G_{y_o}^{\Omega}$ at x_o (condition independant of y_o)(See more details in Brelot [16]).

8. Various complements. (indications).

a) The fine topology may be used too, to study the behaviour of superharmonic functions even at regular boundary points. One may define a Dirichlet problem and Perron-Wiener type envelopes with lim inf or sup in the fine topology at least for relatively compact open sets; that gives the same envelopes (Brelot [13]).

b) Thanks to the notion of uniform integrability introduced in potential theory by Doob [1], we may prove that a harmonic function u in our relatively compact domain ω, when u possesses fine boundary limits a.e (relative to the harmonic measure), forming a function f, and when u satisfies the uniform integrability condition for the harmonic measures $\rho_{x_o}^{\omega_1}$ (ω_1 relatively compact in ω, domain $\omega_1 \ni x_o$) is the solution H_f^{ω} (see Brelot [22]).

c) Let us mention the notions of "thinness of order φ "(Brelot [5]) and of interior thinness (i.e. for any closed set of $e \cup \{x_o\}$) (Brelot [8], Cartan [2]).

d) If u and v are superharmonic > 0 on a Green space Ω, $\frac{u}{v}$ (taken as $+\infty$ when undetermined) has a finite fine limit at every point except on a polar set of μ_v-measure zero i.e. on a v-polar set (Chap VIII, § 6). This result of Doob [4] which contains th. IX, 15) was actually proved as a consequence of a general result of Doob on the behaviour at the Martin boundary (See Chap XVI,§ 4).

e) Fine topology has an important role too in the theory of the so called BLD functions (or BL-made precise functions)[39]. For instance, in any open $\omega \subset \mathbb{R}^n$, a BLD function is fine continuous q.e. (Deny [3]). It does not seem that theorems IX, 12, IX, 15, have been extended to these functions in the general case, but let us mention that a BLD function in $\mathbb{R}^n (n \geqslant 3)$ outside a compact set, has a fine limit at the point at infinity (Deny-Lions).

9. Application to function theory.

A classical Weirstrass theorem says that for a meromorphic function $f(Z)$, in a neighbourhood of Z_0 outside Z_0, the cluster set at Z_0 is one valued (the limit at Z_0) or the extended plane. Doob [7] remarked that the same is true with the fine cluster sets and that precises the second previous case. The proof is immediate and Toda [1] observed that it is valid when Z_0 is replaced by a closed set thin at Z_0. In other words:

Theorem IX, 16. If $f(Z)$ is meromorphic on ω open and $\complement\omega$ is thin at Z_0, the fine cluster set at Z_0 is one-valued or the extended plane.

Proof. If λ is not a cluster value, we may (by a homographic transform) come back to the case of $\lambda = \infty$. Then for a fine neighbourhood σ of Z_0, $f(Z)$ is bounded on $\omega \cap \sigma$. The real and imaginary parts of f have a fine limit at Z_0 (Th. IX, 12), hence also $f(Z)$.

The question has been deepened by Doob [7], then chiefly by Toda [1,2]. For instance: if Z_0 is an isolated boundary point, when there is no limit at Z_0 (essential singularity), but a fine limit, there are no Picard exceptional values (Doob); in the general case, if there is no fine limit at Z_0, the exceptional values for any fine neighbourhood of Z_0 form a locally polar set (Toda).

---

(39) See Deny [3], Deny-Lions [1], Brelot [15], Doob [6]. A general BLD function on an open set $\Omega \subset \mathbb{R}^n$ is real, q.e. finite, and limit q.e. of a sequence of smooth (even C^∞) functions with finite Dirichlet semi-norm (i.e. finite Dirichlet integral), the sequence being a Cauchy sequence for this semi-norm.

CHAPTER X

Relations with the Choquet boundary

1. General Choquet and Šilov boundaries.

Definition X, 1. On a compact space E, let us consider a family $\mathcal{E}$ of extended real valued lower-semi-continuous functions $> -\infty$. Following Choquet, a point X of E is said to be $\mathcal{E}$-extreme if any unitary measure $\mu \geqslant 0$ on E satisfying $\int f(x)\, d\mu(x) \leq f(X)\ \forall f \in \mathcal{E}$, is ε_X (unit mass at X). The set of these points is called the Choquet boundary(39) of E (relative to $\mathcal{E}$).

In case $\mathcal{E}$ is a vector space of real continuous functions, the definition means that for any positive unitary measure μ, the condition $\int f d\mu = f(X)$ for every f implies $\mu = \varepsilon_X$.

When E is a convex compact set in a locally convex Hausdorff topological vector space and $\mathcal{E}$, the set of all continuous linear forms restricted to E, $\mathcal{E}$-extreme points are the usual extreme points of E.

When there exists a smallest compact set where every function of $\mathcal{E}$ attains its minimum, it is called the Šilov boundary of E.

Bauer [1] proved that when the functions of $\mathcal{E}$ separate E and satisfy: $u \in \mathcal{E},\ v \in \mathcal{E} \Rightarrow u + v \in \mathcal{E}$, then the Šilov boundary exists and is the closure of the Choquet boundary. These boundaries are very important. Hence, the interest of their relations with thinness, is as follows.

2. Theorem X, 2 (Bauer [1]). Consider in a Green space Ω_0, a relatively compact open set Ω and the family $\mathcal{E}$ of the real continuous functions on $\bar{\Omega}$ which are harmonic on Ω. Then the Choquet boundary of $\bar{\Omega}$ for $\mathcal{E}$ is the set of the regular boundary points of Ω. (its closure, called sometimes the reduced boundary(40) is the corresponding Šilov boundary).

Proof. That is an easy consequence of the following Keldych lemma (See reference and a simple proof in Brelot [20 bis]).

---

(39) called by Choquet fine boundary as a refinement of the Šilov boundary (See below).

(40) Originally defined as the set of points whose every neighbourhood on $\partial\Omega$ is not locally polar.

If x_0 is a regularboundary point, there exists a function of $\mathcal{E}$ which is zero at x_0 and is > 0 everywhere else.

It is sufficient to use the existence for a given $\varepsilon > 0$, $K > 0$, a neighbourhood δ of x_0, of a function of $\mathcal{E}, \geq 0$, $< \varepsilon$ at x_0 and $> K$ outside δ (lemma easier to prove). Let us use this weakened lemma. Suppose x_0 is regular and μ is a ≥ 0 unitary measure such that $\int f \, d\mu = f(x_0)$ $\forall f \in \mathcal{E}$. Using the function f of the lemma, we get

$$\varepsilon \geq \int f \, d\mu \geq \int_{\complement\delta} f \, d\mu \geq K_1 \mu(\complement\delta).$$

Hence $\mu(\complement\delta)$ is arbitrary small, i.e. $\mu = \varepsilon_{x_0}$ or x_0 is $\mathcal{E}$-extreme.

Conversely, suppose x_0 is $\mathcal{E}$-extreme. Then, $x_0 \notin \Omega$. If not, $\int f \, d\sigma = f(x_0)$ $\forall f \in \mathcal{E}$, where σ is the unitary uniform measure on any small sphere with centre x_0.

Suppose then $x_0 \in \partial\Omega$ and introduce the harmonic measure ρ_x^{Ω} at $x \in \Omega$ and the representation $H_{\varphi}^{\Omega}(x) = \int \varphi(y) \, d\rho_x^{\Omega}(y)$ (φ finite continuous on $\partial\Omega$). Consider the filter $\mathfrak{F}$, trace on Ω of the filter of the neighbourhoods of x_0, and any possible vague limit μ of ρ_x^{Ω} according to a filter $\mathfrak{F}'$ finer than $\mathfrak{F}$.

Then $\int f \, d\rho_x^{\Omega} \xrightarrow[\mathfrak{F}']{} \int f \, d\mu$ $\forall$ $f \in \mathcal{E}$.

But $\int f \, d\rho_x^{\Omega} = f(x) \xrightarrow[\mathfrak{F}]{} f(x_0)$.

Hence, $\int f \, d\mu = f(x_0)$.

As x_0 is $\mathcal{E}$-extreme, $\mu = \varepsilon_{x_0}$. Hence ρ_x^{Ω} converges vaguely to ε_{x_0}, according to $\mathfrak{F}$. Therefore, $H_{\varphi}^{\Omega}(x) \xrightarrow[\mathfrak{F}]{} \varphi(x_0)$. We conclude therefore that x_0 is regular.

<u>Remark</u>. The converse may also be deduced from the existence of a fine limit at x_0 irregular for H_f^{Ω} ($\forall f \in \mathcal{E}$); it is expressed by $\int f \, d\nu_{x_0}$ (ν_{x_0}[41] unitary > 0

---

(41) Actually ν_{x_0} is $b_{\varepsilon_{x_0}}^{\complement\Omega} \neq \varepsilon_{x_0}$ (balayage made in some Greenian domain $\Omega_0 \supset \bar{\Omega}$).

and $\neq \varepsilon_{x_0}$) and must be equal to $f(x_0)$ (continuity of f). That implies that x_0 is not $\mathcal{E}$-extreme.

Exercise. Instead of $\bar{\Omega}$, consider a compact K and the set $\mathcal{E}$ of the real continuous functions on K, harmonic on $\overset{o}{K}$. Show that the Choquet boundary is formed by the regular boundary points of $\overset{o}{K}$ and the points of $K - \overline{\overset{o}{K}}$.

Theorem X, 3. In a Green space Ω_o, consider a compact K and the set $\mathcal{E}$ of functions on K such that every one is the restriction on K of a harmonic function on an open neighbourhood of K. Then the $\mathcal{E}$-extreme points are the stable boundary points of K, the Choquet boundary of K for $\mathcal{E}$ is the fine boundary of K and the Šilov boundary is ∂K(42).

Proof. As previously, no interior point is $\mathcal{E}$-extreme.

Now, the solution of the Dirichlet problem for K and φ finite continuous on ∂K (Chap VI, § 6 δ) is denoted K_φ. $\varphi \rightsquigarrow K_\varphi(x)$ is an increasing linear functional, therefore expressed by $\int \varphi \, d\nu_x$ (ν_x positive unitary measure)(43) and $\nu_x = \varepsilon_x$ iff $x \in \partial K$ is stable. Note that $\forall f \in \mathcal{E}$, $K_f = f$ on K. There fore, if $x \in \partial K$ is unstable, $f(x) = \int f \, d\nu_x \; \forall f \in \mathcal{E}$, with $\nu_x \neq \varepsilon_x$ and x is not $\mathcal{E}$-extreme.

Conversely, if $x_0 \in \partial K$ is not $\mathcal{E}$-extreme, there exists a unitary measure $\rho_{x_0} \geqslant 0$, $\neq \varepsilon_{x_0}$ such that $f(x_0) = \int f \, d\rho_{x_0} \; \forall f \in \mathcal{E}$. Let U be subharmonic $\geqslant 0$ zero at x_0 only (like $|x - x_0|$ in $\mathbb{R}^n$)(44). Then, for an open neighbourhood ω of K,

$$H_U^{\omega}(x_0) = \int H_U^{\omega}(x) \, d\rho_{x_0}(x) \geqslant \int U \, d\rho_{x_0} > 0.$$

---

(42) The condition of separation of Bauer (§ 1) is easy in $\mathbb{R}^n$ and is satisfied in general by considering $x \to G_y^{\Omega}(x)$ for $y \in \complement K$ and using the symmetry of G and the analyticity of harmonic functions

(43) Actually ν_x is $b_{\varepsilon_x}^{\complement K}$ and is identical to ε_x iff $\complement K$ is not thin at x (i.e. x is stable).

(44) If x_0 is not polar, $G_{x_0}(x_0) - G_{x_0}(x)$ solves the question. If not, it is sufficient to consider $u_n = \mathrm{Inf}\,(G_{x_0}\, n)$ and $\sum \lambda_n u_n(x_0) - \sum \lambda_n u_n(x)$ with a $(\lambda_n) > 0$ making finite the first sum.

Hence $K_U(x_0) > 0$; x_0 is unstable i.e. $\complement K$ is thin at x_0.

We may complete by recalling that the set of the stable points is dense on ∂K.

<u>Proof</u>. If not, $\complement K$ is thin on a set $e = \sigma \cap \partial K$ for an open set σ; then, for every component of $\Omega_1 - K$ (Ω_1 open, $\Omega_1 \supset K$, $\bar{\Omega}_1 \subset \Omega_0$), e would be of harmonic measure zero and there would exist on $\Omega_1 - K$ a superharmonic > 0 function tending to $+\infty$ at the points of e. The continuation by $+\infty$ on K would be superharmonic on σ, and $\sigma \cap K$ would be locally polar, therefore $\complement K$ is not thin at the points of e - a contradiction.

<u>Remark</u>. The families $\mathfrak{E}$ are particular cases of a general axiomatic study of Bauer [1] who considers on a compact space, a convex vector space of real continuous functions containing the constants and separating the points of the compact space; a corresponding Dirichlet problem is defined with the Šilov boundary.

<u>Exercise</u>. In the previous theorms, determine more directly the Šilov boundaries.

CHAPTER XI

Extension to axiomatic theories of harmonic functions

(Short indications)

1. We will give only some notions in some important cases.

By modifying the starting points of a theory of Doob [1], Brelot [19, 20][(43)] developed an extension of the classical theory of harmonic and superharmonic functions in the following way.

Given a locally compact non-compact locally connected Hausdorff space Ω, to every open set ω, a vector space of real continuous functions (called harmonic functions in ω) is associated. $\bar{\Omega}$ Will be the Alexandroff compactification of Ω; we will use its topology.

Axiom 1 (Sheaf axiom). Any harmonic function in ω is harmonic in any ω' open $\subset \omega$ and any function, harmonic in a neighbourhood of every point of ω is harmonic in ω.

Regular set. An open ω is regular if $\bar{\omega} \subset \Omega$ and if any real continuous function f on $\partial\omega$ has a unique continuous continuation H_f on $\bar{\omega}$, harmonic in ω and increasing with f; it is therefore a linear form $\int f \, d\rho_x^\omega$ ($\rho_x^\omega \geqslant 0$, called harmonic measure at x).

Axiom 2. The regular open sets form a base of the topology of Ω.

Axiom 3. On a domain ω, any increasing sequence of harmonic functions tends to a harmonic function or to $+\infty$.

According to Constantinescu-Cornea, this implies the same for a directed increasing set and thanks to Mokobodzki-Loeb-Walsh, this axiom 3 (thanks to the other axioms) is equivalent to the property that any family of harmonic functions

---

(43) Most of the further developments of various authors, partly used in the text, are indicated in Brelot [28].

$u_1 \geqslant 0$ on a domain ω and upper bounded at a point is equicontinuous at every point.

2. Definition. u is hyperharmonic on an open set ω_0 if u is lower-semi-continuous, greater than $> -\infty$ and majorizes $\int u \, d\rho_x^{\omega}$, for any regular $\omega \subset \overline{\omega} \subset \omega_0$.

If ω_0 is connected, such a u is either everywhere $+\infty$, or else is finite on a dense set and is then called superharmonic. On an open ω_0, u hyperharmonic is called superharmonic when finite on a dense set.

Minimum Principle: a) a hyperharmonic function $u \geqslant 0$ on a domain is everywhere zero or everywhere > 0.

b) If there exists in an open set ω a superharmonic function $u \geqslant \varepsilon > 0$, any hyperharmonic function u on ω is $\geqslant 0$ when lim inf $u \geqslant 0$ at every boundary point. In case the constants are harmonic, for any u on any ω

$$\inf_{y \in \omega} u(y) = \inf_{x \in \partial\omega} (\liminf u \text{ at } x \in \partial\omega).$$

It is often possible to come back to this case, thanks to the following definition and remark:

h-harmonic functions: If h is a real continuous function > 0 on Ω, the quotients $\frac{u}{h}$ (u harmonic on an open set) define a new sheaf satisfying the axioms (with the same regular open sets). The corresponding hyperharmonic functions (called h-hyperharmonic functions) are the quotients by h of the previous ones.

In case h is harmonic in the given sheaf, the h-harmonic functions contain the constants.

Examples. The solutions of a partial differential equation of elliptic type (of second order) on a domain of $\mathbb{R}^n$ (with sufficiently smooth coefficients) satisfy the axioms and suitable extensions were made to discontinuous coefficients. (Mrs. Hervé [1, 2, 3]).

Potentials. A superharmonic function majorizing a harmonic function, in an open set ω, has a greatest harmonic minorant; when this one is zero, v is called a potential.

The existence in Ω of a potential $V > 0$ is equivalent to the existence in Ω of a superharmonic function > 0, which is not harmonic and is satisfied when there exist two non-proportional, strictly positive harmonic functions (therefore is satisfied locally). Ω is then generalising the Green space.

Denote by (A), the set of hypotheses 1,2,3, existence of a potential >0 and (A_1) the same with a countable base.

Consequence of (A). a) Given any point x there exist potentials with support x (i.e. harmonic on $\complement\{x\}$)[44] but they need not always be proportional. The case of "proportionality" will be important (case denoted A P, and A_1P with a countable base).

b) The non-negative hyperharmonic functions on Ω define a cone Φ of the previous general theory. The locally polar sets are polar and some first properties of classical thinness may be extended without further hypothesis, like hyperthinness and strong thinness for thinness at any polar point $x_0 \notin e$, existence of a fine limit of any superharmonic function $\geqslant 0$ on an open set ω at a boundary point where $\complement\omega$ is thin.

In the case of A_1, weak thinness[45] is equivalent to the thinness at any $x \notin e$; with more hypothesis, we get further extensions[46].

Dirichlet problem. First with (A), it may be developed with the basic topology at least for a relatively compact open set ω ; there is resolutivity of any real continuous boundary function. A regular boundary point x_0 (defined like in the classical case) is characterized by the weak thinness of $\complement\omega$ at x_0.

With (A_1) the resolutivity theorem holds as in the classical case.

---

(44) Constantinescu-Cornea [2] I. Previously with A_1 see Mrs. Hervé [1].

(45) Actually the weak thinness of $e \not\ni x_0$ at x_0 is equivalent to the thinness when there exists a countable base of neighbourhoods of x_0, therefore always with A_1. This is a consequence of further researches of Bauer and Constantinescu- Cornea (see ref. in Brelot [28] Chap II).

(46) See Brelot [19,20,33], Mrs. Hervé [1]; about the behaviour of bounded harmonic functions on an open ω in the neighbourhood of a boundary point x_0 where $\complement\omega$ is thin, see Smyrnelis [1].

3. Extension of the Riesz representation (with A_1).

If we order the superharmonic functions by the condition

$$u_1 \succ u_2 \Longleftrightarrow u_1 = u_2 + \text{a superharmonic function} \geqslant 0$$

(which defines the so called specific order), the cone S^+ of the non-negative superharmonic functions is a lattice (even complete). With the following equivalence relation of pairs of non-negative superharmonic functions,

$$((u_1, u_2) \backsim (u_1', u_2')) \Longleftrightarrow (u_1 + u_2' = u_2 + u_1'),$$

let us consider the corresponding equivalence classes $[u_1, u_2]$ whose set defines a vector space S. With a suitable topology T on S (Mrs. Hervé)[(47)], we get a locally convex Hausdorff topological vector space where S^+, identified with the set of all $[u, 0]$, has compact metrizable bases; then the classical Choquet theorem on extreme elements gives for any u (superharmonic $\geqslant 0$) with such a base B.

$$l(u) = \int l(v)\, d\mu\,(v),\ v \in B\ (l \text{ any linear continuous form}),$$

from where $u(x) = \int v(x)\, d\mu(v)$ where μ is a unique positive unitary measure on B, charging only the set of extreme points of B. These extreme points are either harmonic (set H) or potentials with point support (set P). Hence $u(x) = \int v(x)\, d\mu_1(v) + \int v(x)\, d\mu_2\,(v)$, μ_1 charging only H and μ_2 charging only P. The first part is the greatest harmonic minorant of u; in the case of proportionality, P is homeomorphic to Ω, the second part may be written $\int p_X(x)\, d\mu_2'\,(X)$ (where μ_2' on Ω corresponds to μ_2 on B, charging only P)

---

(47) With a new axiom (existence of a base β of "completely determining" domains δ, see § 5) the semi-norms of (u_1, u_2) equal to $\left|\int u_1 d\rho_x^\delta - \int u_2\, d\rho_x^\delta\right|$ ($\delta \in \beta$, $x \in \delta$) defines a topology on S which furnishes S^+ with a compact metrizable base (inspired by Cartan in the classical case (unpublished); developed in Brelot [19]). Mrs. Hervé [1] avoided the new axiom and Mokobodzki gave another simpler introduction of this topology (See Mokobodzki [1], Brelot [28]).

and the first part leads to the introduction of a general Martin boundary (see later, Part II).

4. Axiom D. The theory will extend many refinements of the classical one thanks to the following new axiom: Any locally bounded potential v on Ω, harmonic in an open set ω (for example the greatest open set, whose complement is called the support of v) is majorized by any non-negative superharmonic function u which is $\geqslant v$ on $\complement\omega$.

That implies the same property for any subdomain of Ω and there is equivalence with the local property in the case of a countable base in Ω).

Some consequences of $A_1 + D$.

a) A convergence theorem holds for superharmonic functions, like in the classical case and it is even equivalent to axiom D in the case of proportionality.

b) Weak thinness $\Longleftrightarrow$ thinness $\Longleftrightarrow$ strong thinness and semi-polar sets are polar.

c) Unthinness of e at $x_0 \notin e$ is always strong (even without a countable base).

d) The points of thinness of e on e form a polar set and even a Choquet property holds, for instance for the weight $\int \hat{R}_1^e \, d\rho_{x_0}^{\omega_0}$ and that is important for extensions of the behaviour of various capacities of decreasing sets, where fine topology is essential as in the classical case (see Brelot [26 to 30]).

5. Adjoint Sheaf. (Mrs. Hervé [1]). We use the set of hypothesis (A_2) formed by A_1 P and the existence of a base of relatively compact open sets which are "completely determining" i.e. such that for any potential v, harmonic on δ, $R_v^{\complement\delta} = v$ on δ.

Let us choose p_y, potential of support $\{y\}$, for instance by the condition that it is in a fixed compact base of S^+ (with Mrs. Hervé's topology). Then any potential may be written $\int p_y(x)\, d\mu(y)$ where μ is a unique non-negative measure on Ω.

Now, for any relatively compact open set ω, $\hat{R}_{p_y}^{\complement\omega}$ is a potential, therefore has a representation $\int p_z(x)\, d\sigma_y^{\omega}(z)$. Consider the completely determining

$\delta \subset \bar{\delta} \subset \omega_0$ and the finite continuous f on ω_0 satisfying $f(y) = \int f(z) d\sigma_y^{\delta}(z)$. These f define, for all ω_0, a sheaf satisfying the axioms 1,2,3 with the δ's for a base of regular open sets and σ_y^{δ} for harmonic measure. A change in the choice of p_y multiplies the adjoint harmonic functions by a real positive continuous function.

Now $y \rightsquigarrow p_y(x)$ is a corresponding potential denoted $p_x^*(y)$ with support $\{x\}$.

__Example.__ Considering in $\Omega \subseteq \mathbb{R}^n$, a partial differential equation of 2nd order and of elliptic type with locally lipschitzian coefficients, and the solutions, a suitable choice of p_y gives an adjoint sheaf which corresponds to the classical adjoint equation (Mrs. Hervé [1]).

6. __Balayage.__ Only with (A) and even with weaker hypotheses (Boboc-Constantinescu-Cornea), one may extend important properties of R_{φ}^e and chiefly R_u^e, $\hat{R}_u^e$ for superharmonic $u \geqslant 0$ (additivity in u, strong subaddivity in e etc). Note also that with $A_1 + D$, $\hat{R}_u^e$ is the smallest superharmonic function $\geqslant 0$, majorizing u on e quasi-everywhere. The use of fine topology is precious.

As for the balayage of a measure, the theory will differ seriously from the classical one, by lacking of a symmetric Green function. Starting from (A) consider a measure $\mu \geqslant 0$ with compact support; the equality $\int u \, d\, b_{\mu}^e = \int \hat{R}_u^e \, d\mu$, inspired by the classical case, supposed to hold for any real continuous potential u, defines a unique measure $b_{\mu}^e \geqslant 0$ (called the balayaged measure). With additional hypotheses, the formula may be extended to general superharmonic functions u.

The theory has many applications; already with (A_1), a criterion of __thinness__ of e at x_0 is $b_{\varepsilon_{x_0}}^e \neq \varepsilon_{x_0}$ (Constantinescu [2]).

With A_2, consider an adjoint sheaf (§ 5). If an asterisc means an adjoint notion, a key-formula (Mrs. Hervé) is $\hat{R}_{p_y}^e(x) = \hat{R}_{p_x^*}^{*e}(y)$ which implies __for instance__.

a) identity of adjoint polar sets and polar sets.

b) equivalence of axiom D for the original sheaf and the adjoint one.

c) criterion of __thinness__ of e at $x_0 \notin e$: for a neighbourhood δ of x_0

$R^{*e\cap\delta}_{p^*_{x_0}} \not\leq p^*_{x_0}$ or when x_0 is polar (even when it is on e) $\hat{R}^{*e}_{p^*_{x_0}} \not\leq p^*_{x_0}$.

But we cannot detail here the rich extensions of the classical case with various hypotheses. (See Mrs. Hervé [1], Boboc-Constantinescu-Cornea [1,2], Constantinescu-Cornea [2], Constantinescu [2], Brelot [27, 28, 30, 32, 33].

7. Dirichlet problem for compact sets.

Extension of the classical case has been made. De la Pradelle [1,2], first with A_1, then with many more hypotheses, shows the role of thinness and also the connections with a property of quasi-analyticity (i.e. a harmonic function is zero when it is zero in a neighbourhood of a point).

Let us note also that the relations with the Choquet boundary may also be adapted to the previous axiomatic, (with more or less hypotheses) but it does not seem it has been published.

8. Weaker axiomatics. In order to get applications to equations of parabolic type, Bauer [2,3,5] weakened the previous axioms as follows. Upto details, he keeps the axioms 1,2 and weakens the axiom 3 by saying that for an increasing directed family of harmonic functions u_i on an open set ω, sup u_i is harmonic: (K_1) when sup u_i is bounded or (K_2) when sup u_i is finite or (K_D)(Doob's axiom) when sup u_i is finite on a dense set. In order to get a minimum principle, a further axiom is introduced: (T) There exists a harmonic function $h > 0$ on Ω, and the h-hyper harmonic functions (same meaning as in the classical case) separate Ω (often the positive ones are supposed to separate; that is T^+).

Note that these axioms are satisfied in the first axiomatic, when there exist two non-proportional harmonic functions in Ω.

With a countable base and such axioms, generally K_D and the existence for every $x \in \Omega$ of a potential > 0 at x, Bauer develops a theory similar to the previous one. However an integral representation is more difficult and not so useful because there is no base of S^+. Thinness has a similar role.

The advantage is the applications also to the heat equation and to a large class of parabolic equations of second order. Boboc-Constantinescu-Cornea[2]),

B. Walsh etc. weakened more these axioms and larger theories of local or global character and converse questions are being developed by Hansen and chiefly Mokobodzki and Sibony (see Mokobodzki [2]).

About topologies in these axiomatics, let us mention, although outside our frame, the important role of nuclearity introduced by P. Loeb-B. Walsh (see also Hinrichsen [1]).

As for the applications of these axiomatics, we had to determine all the sheaves satisfying the axioms. The case of a domain of $\mathbb{R}^n$ with C^2-functions or of the whole $\mathbb{R}^n$ and sheaves invariant by translation, with constants harmonic, has been clarified by Bony [1] even with axioms 1,2 alone, with a corresponding pre-elliptic (possibly degenerated) operator. General researches with not so smooth functions (like the solutions of equations with discontinuous coefficients studied by Stampacchia [1] and considered in the previous frame by Mrs. Hervé [2, 3] remain open.

CHAPTER XII

Abstract Minimal thinness, Minimal boundary, Minimal fine topology

1. Introduction to Part II

The boundary problems for Partial Differential Equations were studied formerly only with Euclidean boundaries. A small improvement was the introduction of the point at infinity of $\mathbb{R}^n$ and the use of the compactification. In classical projective geometry, the points, lines, plane at infinity were used too.

A first example of another important boundary was given by the prime ends of Caratheodory, for the study of the behaviour of the functions of a complex variable. Now this appears as a particular case of the Martin boundary (see Chap XIV), compactification or completion. Specialists of the Dirichlet problem were led also to decompose certain ordinary boundary points: it is obvious that for a disc minus a radius, there is no reason why a wanted limit at a point of the radius must be the same for both ways of approach. In order to give a better form to such considerations, it was natural to introduce suitable metrics on the domain, compatible with the Euclidean topology and to complete (see Brelot [9]).

But in order to get more results, it soon appeared it was better to introduce boundaries depending on the nature of the functions to be studied. Consider even a general topological space Ω and functions which take their values in some other space Ω', useful boundaries of Ω may be obtained by compactification or completion depending on the set of these functions. Such considerations will be developed in the next chapter with further applications.

Now the behaviour of the considered functions may also sometimes be studied in a natural way, according to certain sequences of points or to certain filters. Such filters may be considered as points of an abstract boundary and it may be interesting, even if not necessary, to introduce on the union of the space Ω and of this abstract boundary, a topology for which limits according to the considered filters are ordinary limits in this topology. We shall develop also such ideas, but by choosing the filters in connection with the notion of extreme elements, which is more and more important in analysis (Choquet's theory); that will include the "minimal boundary", "minimal thinness" at points of the boundary and a continuation of a previous fine topology, called "minimal fine topology". Naim-Doob's classical results on harmonic and superharmonic functions, extended to axiomatic harmonic functions by Gowrisankaran, and giving essential applications to functions of a complex variable or to Partial differential equations show the fundamental importance of these notions. In large cases, we will see also later that this minimal fine topology may be interpreted in the frame of Part I.

2. Preliminaries: (based on Myskis' study [1]).

Theorem XII, 1. Let Ω be a set with a given topology making a space Ω_0. We consider on Ω_0 a family $\{\mathcal{B}_i\}$ $(i \in I)$ of bases of filters, whose sets are open and the set $\Omega_1 = \Omega \cup I$. There exist on Ω_1 topologies such that

1) the induced topology on Ω is the topology of Ω_0.

2) the neighbourhoods of any $i \in I$ intersect Ω according to sets which form a filter with base $\mathcal{B}_i$.

There exists a finest one T_M; it leaves Ω open and induces on I, the discrete topology; the sets of $\mathcal{B}_i$ with added i form a base of neighbourhoods of i.

Among the previous topologies which make Ω open, there exists a coarsest one T_m; the neighbourhoods of $x_0 \in \Omega$ in Ω_0 form a base of neighbourhoods in T_m (as well as in T_M); as for $i \in I$, we associate a set $\alpha \in \mathcal{B}_i$ and the set J of the points $j \in I$ such that α contains a set of $\mathcal{B}_j$. Then, when α varies in $\mathcal{B}_i$, $\alpha \cup J_\alpha$ describes a base of a filter which is the filter of neighbourhoods of i in Ω_1 according to T_m.

Finally T_M is Hausdorff iff:

a) Ω_0 is Hausdorff

b) $\forall\, x \in \Omega_0$, $i \in I$, there exists a neighbourhood of x in Ω_0 and $\alpha \in \mathcal{B}_i$ which are disjoint.

c) $\forall\, i, j \in I$, $i \neq j$, there exist disjoint α, β, $\alpha \in \mathcal{B}_i$, $\beta \in \mathcal{B}_j$.

__Proof.__ Let us consider on Ω_1 the topology defined by the neighbourhoods of $x \in \Omega_0$ as forming a base of neighbourhoods of x and by $\alpha \cup \{i\}$, α describing $\mathcal{B}_i$, as giving a base of neighbourhoods of i. The axioms of neighbourhoods are actually satisfied; this topology satisfies the required conditions 1,2; it makes Ω open and induces on I the discrete topology.

In any topology satisfying conditions 1,2, a neighbourhood of $x \in \Omega$ or $i \in I$ is obviously a neighbourhood of the previous topology which is therefore the finest one T_M.

Consider now the question relative to T_m. Let us check that the corresponding class of sets indicated above satisfy the axioms of neighbourhoods. It is obvious for any $x \in \Omega$; consider an i, a corresponding $\alpha \cup J_\alpha$ and let us see it is a proposed neighbourhood (in the previous sense denoted $\mathcal{S}$) of everyone of its points. Obvious for a point of α (open in Ω_0); as for a $j \in J_\alpha$, there exists a $\beta \in \mathcal{B}_j$ contained in α. Any j' such that $\mathcal{B}_{j'}$ has an element $\beta' \subset \beta$, is in J_α. Therefore β and these j' form a $\mathcal{S}$-neighbourhood contained in $\alpha \cup J_\alpha$. We conclude that the axioms are satisfied and this topology satisfies conditions 1,2 and makes Ω open.

It remains to compare $\mathcal{S}$ with any topology T_1 satisfying 1,2 and making Ω open. Consider any $\mathcal{S}$-neighbourhood $\mathcal{V}$ of a point of Ω_1 and let us prove it is a T_1-neighbourhood. Obvious if the point is in Ω. If it is a $i \in I$, $\mathcal{V}$ contains a $\alpha \cup J_\alpha$; α is the intersection with Ω of a T_1-neighbourhood of i whose T_1-interior is α' (say); if $j \in I$, $j \in \alpha'$, then α' is a T_1-neighbourhood of j and $\alpha' \cap \Omega$ contains an element of $\mathcal{B}_j$; but $\alpha = \alpha' \cap \Omega$, therefore $j \in J_\alpha$ and $\alpha' \subset \alpha \cup J_\alpha \subset \mathcal{V}$; $\mathcal{V}$ is therefore a T_1-neighbourhood of i. We conclude therefore that T_1 is finer than $\mathcal{S}$.

The last properties are obvious.

__Remark.__ An interesting case where the topologies satisfying (1) and (2) are making Ω open is the case where, $\forall\, x_0 \in \Omega$, there exists in Ω a neighbourhood which is never in the filter of any base $\mathcal{B}_i$.

3. Abstract minimal thinness.

A little more generally than like Gowrisankaran [1], we consider on a non-void set Ω two families of non-negative functions.

One is a cone U of real-valued functions (called harmonic) containing the function zero.

One is a convex cone P of extended real-valued functions (called potentials) i.e. $p_1 \in P$, $p_2 \in P \Rightarrow \alpha_1 p_1 + \alpha_2 p_2 \in P$ (α_1, α_2 real $\geqslant 0$) (convention $0.\infty = 0$).

We denote by Σ the cone $U + P$ of the functions $u + p$ ($u \in U$, $p \in P$). We shall suppose:

Axiom $\mathcal{A}_1$. $u \in U$, $p \in P$, $u \leqslant p \Rightarrow u = 0$.

Axiom $\mathcal{A}_2$. If $u \in U$, $v \in \Sigma$, then

$$\inf (u, v) \in \Sigma .$$

In case there exists a $p = +\infty$ everywhere, $\mathcal{A}_2$ implies $u = 0$ ($\forall u \in U$).

Minimal harmonic functions.

Definition XII, 2. $h \in U$ will be said minimal if $u \in U$, $u \leqslant h$ implies $u = \alpha h$ ($0 \leq \alpha$ real).

Important particular case $\mathcal{C}$. Suppose U is convex, consider the vector space $U - U$. Suppose that with the natural order of the functions, the elements greater or equal to zero are the elements of U. i.e. U is the positive cone of this vector space. Then for $h \in U$, $h \neq 0$

$$h \text{ minimal} \Longleftrightarrow \{\lambda h\} \ (\lambda \geqslant 0) \text{ is an extreme generatrix of the cone } U.$$

Reduced function. Like previously, consider $E \subset \Omega$, f a real $\geqslant 0$ function on E, majorized by a function of Σ. Then we define

$$R_f^E = \inf_{v \in \Sigma} v \ , \quad R_f^{\Omega} \text{ is written } R_f .$$

Theorem XII, 3. For $h \neq 0$, minimal harmonic and $E \subset \Omega$, the following properties are equivalent.

a) $R_h^E \not\equiv h$ i.e. <u>there exists a</u> $v \in \Sigma_1$, <u>such that</u> $v \geqslant h$ <u>on</u> E, <u>but not everywhere</u>.

b) <u>There exists a potential majorizing</u> h <u>on</u> E.

c) $\forall u \in U$, the condition $u \leqslant R_h^E$ implies $u = 0$.

<u>Proof</u>. Starting from (a) and the considered function v, we see that $w = \inf(v,h)$ satisfies $w \in \Sigma_1$, $w \leqslant h$, $w = h$ on E, $w \not\equiv h$,

Now, a decomposition $w = u + p$ gives $u \leqslant h$, hence $u = \alpha h$ where $0 \leq \alpha < 1$, therefore $\alpha h + p = h$ on E, $\frac{p}{1-\alpha} = h$ on E. We get (b).

Starting from (b), a potential majorizing h on E, majorizes R_h^E. Hence (c). Finally (c) implies that $h \not\leqslant R_h^E$, hence $h \not\equiv R_h^E$, i.e. (a)

<u>Definition XII, 4</u>. $E \subset \Omega$ is said to be <u>thin</u> relatively to h (minimal $\not\equiv 0$) if $R_h^E \not\equiv h$ (or satisfies any of the previous equivalent conditions).

This definition (Gowrisankaran [1]) is inspired by the classical concept of L. Naïm [1] and by the previous form (b) in the classical case of a half-plane for the so-called P.L sets of Ahlfors-Heins [1].

<u>Remarks</u>. 1) Ω is never thin ($\forall h$); ϕ is always thin; a subset of a thin set is thin.

2) $\{x \mid h(x) = 0\}$ is thin relatively to h (minimal $\not\equiv 0$).

Because the potential 0 majorizes h on this set.

<u>Theorem XII, 5</u>. <u>Let</u> h <u>be minimal harmonic</u>, $\not\equiv 0$. <u>The union of</u> e_1 <u>and</u> e_2, <u>thin relatively to</u> h, <u>is thin i.e., the complementary sets of the thin sets form a filter</u> $\mathcal{F}_h$.

<u>Proof</u>. If e_1, e_2 are thin and p_1, p_2 are potentials majorizing h on e_1, e_2, $p_1 + p_2$ is a potential majorizing h on $e_1 \cup e_2$.

<u>Theorem XII, 6</u>. <u>Let</u> h <u>be minimal</u> $\not\equiv 0$, <u>and</u> $v \in \Sigma_1$. <u>On the set</u> Λ <u>where</u> $\frac{v}{h}$ <u>has a meaning</u>, $\frac{v}{h}$ <u>has a limit according to</u> $\mathcal{F}_h$ <u>and this limit is finite and is equal to</u> $\inf_{\Lambda} \frac{v}{h}$. (Naïm [1] Th. 8,17, Gowrisankaran [1]).

<u>Proof</u>. As $\complement\Lambda \subset \{x \mid h(x) = 0\} = e$, $\complement\Lambda$ is thin. Denote $\alpha = \inf_{\Lambda} \frac{v}{h}$; α is finite; if not, v and any potential part p of a decomposition would be $+\infty$ on $\complement e$; hence $h \leqslant p$ everywhere and h would be zero.

Consider now the set $E_\varepsilon = \{x \in \Lambda, \frac{v}{h} \geqslant \alpha + \varepsilon\}$, $\varepsilon > 0$. Obviously $E_\varepsilon \supset e \cap \Lambda$ and $E_\varepsilon \neq \Omega$; $\frac{v}{\alpha+\varepsilon}$ is $\geqslant h$ on $E_\varepsilon \cap \complement e$, but not on $\complement E_\varepsilon$. Hence E_ε is thin, i.e. $\complement E_\varepsilon \in \mathcal{F}_h$ and $\frac{v}{h}$ tends to α according $\mathcal{F}_h$ on Λ.

Corollaries. Let h, h' be minimal $\not\equiv 0$, not proportional.

1) on the set E where $\frac{h'}{h}$ has a meaning $\frac{h'}{h} \xrightarrow[\mathcal{F}_h]{} 0$.

2) There exists in $\mathcal{F}_h$ and $\mathcal{F}_{h'}$ respective disjoint sets.

Proof. (1) is a consequence of the fact that $\frac{h'}{h} \geqslant K > 0$ on E, would imply $h' = \lambda h$ everywhere

(2) is proved by considering the subset of E where $\frac{h'}{h} < 1$ (element of $\mathcal{F}_h$) and the other subset where $\frac{h'}{h} > 1$ (element of $\mathcal{F}_{h'}$).

4. Minimal boundary:

Two minimal h will be said equivalent if they are proportional (with factor $\neq 0$). The equivalence class of h will be denoted $\bar{h}$. $\mathcal{F}_h$ is the same for all h of the same equivalence class $\bar{h}$ and we will write $\mathcal{F}_{\bar{h}}$ for all these identical filters.

Definition XII, 7. These classes h will be called minimal boundary points and their set, (abstract) minimal boundary $\mathcal{M}$.

The notions of lim, lim sup... corresponding to $\mathcal{F}_{\bar{h}}$ will be said fine lim ... etc. at the point $\bar{h}$.

Note that in the case $\mathcal{C}$, the $\bar{h}$ correspond to the extreme generatrices of U and when there is a base of U, to the extreme points of such a base.

Topological interpretation.

Suppose Ω is a topological space, U a convex cone of real non-negative continuous functions and P, a convex cone of lower semi-continuous non-negative functions. The functions of $U + P$ and $+\infty$ form a convex cone Φ satisfying the conditions of Chap I. There is a corresponding fine topology $\mathcal{C}$ on Ω.

Now for a continuous function $f \geqslant 0$, $R_f^E = R_f^{\tilde{E}}$ ($\tilde{E}$, the fine closure of E). Therefore, if e is thin at $\bar{h}$, $\tilde{e}$ will be thin too and the complementary sets of the fine closed sets which are thin at $\bar{h}$ form a base of $\mathcal{F}_{\bar{h}}$ whose elements are fine open. We may use theorem XII, 1, where I is now $\mathfrak{M}$. We get immediately,

Theorem XII, 8. There exist on $\Omega \cup \mathfrak{M}$ topologies such that for every one

1) It induces on Ω the fine topology $\mathcal{C}$.

2) The corresponding filter of neighbourhoods of any $\bar{h} \in \mathfrak{M}$ induces on Ω the filter $\mathcal{F}_{\bar{h}}$.

For such a topology, the corresponding notions of lim, lim sup.. at $\bar{h}$ from Ω are the same notions relative to $\mathcal{F}_{\bar{h}}$.

Among these topologies there is a finest one (which makes Ω open and induces on $\mathfrak{M}$ the discrete topology) and among those making Ω open, there is a coarsest one that we may call minimal fine topology (precised in XII, 1).

Note that, thanks to the second corollary, this last topology is Hausdorff when the fine one on Ω is Hausdorff, in case there exists, $\forall x \in \Omega$, a potential p such that $p(x) > 0$ (because this implies for every $\bar{h}$, the thinness of a suitable neighbourhood of x).

We will study important particular cases where this topology on $\Omega \cup \mathfrak{M}$ will be the fine one (meaning of Chap I) corresponding to a suitable family of lower-semi-continuous functions for a suitable topology on $\Omega \cup \mathfrak{M}$.

Exercise. In the hypotheses of the topological interpretations, suppose more:

i) countable addivity in Φ

ii) for h minimal $\neq 0$, $u \leq u'$ ($u, u' \in U$) implies $u - \lambda h \leq u' - \lambda' h$ where $\lambda = \inf_{\Omega} u/h$, $\lambda' = \inf_{\Omega} u'/h$ (where these quotients have meaning)

iii) there exists a sequence $\mathcal{V}_n$ of sets such that, for every set e thin at $\bar{h}$ (same h) $R_h^{e \cap \mathcal{V}_n} \longrightarrow 0$.

Then, given $u_0 \in U$, there exists a set e_0 thin at this $\bar{h}$ such that for any $u \in U$, $u \leq u_0$, u/h has a limit according to a base of filter $\mathcal{V}_{n_p} \setminus e_0$ (n_p a subsequence) (limit equal to the minimal fine limit at $\bar{h}$, i.e. according to $\mathcal{F}_{\bar{h}}$).

CHAPTER XIII

General Compactification of Constantinescu-Cornea

First examples of application

1. A theorem of Constantinescu-Cornea, given for Riemann surfaces [1], but with a proof holding for locally compact spaces and actually close to the Stone-Cech compactification, will allow the introduction of many useful boundaries in a common way.

Theorem XIII, 1. Let Ω be a non-compact locally compact Hausdorff space. Let Φ be a family of continuous functions from Ω to $[-\infty, \infty]$. Then there exists one and only one compact space $\hat{\Omega}$, upto homeomorphism, such that

i) Ω is a dense subset of $\hat{\Omega}$

ii) Every $f \in \Phi$ can be extended to a continuous function $\hat{f}$ on $\hat{\Omega}$

iii) The family of the $\hat{f}$ separates $\hat{\Omega} - \Omega$ (denoted Δ).

Moreover in this space $\hat{\Omega}$, Ω is open.

Proof. Let us see first that Ω is open in any $\hat{\Omega}$ satisfying the previous conditions. Consider all relatively compact open sets Ω_i in the space Ω;

$\Delta \subset \bar{\Omega} \subset (\overline{\Omega - \Omega_i}) \cup \bar{\Omega}_i$ (closures in $\hat{\Omega}$). As the closure of Ω_i in Ω is compact, it is also the closure in $\hat{\Omega}$. Hence $\bar{\Omega}_i \subset \Omega$ and $\Delta \subset (\overline{\Omega \setminus \Omega_i})$, $\Delta \subset \bigcap_i (\overline{\Omega \setminus \Omega_i})$. But $(\overline{\Omega \setminus \Omega_i}) \cap \Omega$ is just $\Omega \setminus \Omega_i$ (closed in Ω). Hence $\bigcap_i (\overline{\Omega \setminus \Omega_i}) \cap \Omega = \emptyset$, $\bigcap_i \overline{\Omega \setminus \Omega_i} \subset \hat{\Omega} - \Omega = \Delta$, $\Delta = \bigcap_i (\overline{\Omega \setminus \Omega_i})$.
Δ is therefore compact, and Ω is open in $\hat{\Omega}$.

Let us prove now the uniqueness of an $\hat{\Omega}$, when it exists. Let $\hat{\Omega}, \hat{\Omega}'$ be two spaces satisfying the conditions. The filter of neighbourhoods of $x \in \hat{\Omega} \setminus \Omega$ in $\hat{\Omega}$, has a trace $\mathcal{F}_\Omega$ on Ω; $\mathcal{F}_\Omega$ (base of filter) must converge in $\hat{\Omega}'$: if not, there would be on Ω two filters finer than $\mathcal{F}_\Omega$ (bases of filter $\mathcal{F}_1, \mathcal{F}_2$ on any set $\supset \Omega$) converging to X_1, X_2 $(X_1 \neq X_2)$ in the space $\hat{\Omega}'$; these points are not in Ω; because the convergence to $X_1 \in \Omega$ of $\mathcal{F}_1$ in $\hat{\Omega}'$ would imply the convergence in Ω, then in $\hat{\Omega}$ and that is not compatible with the convergence of $\mathcal{F}_{\Omega}$ to x in $\hat{\Omega} - \Omega$. Starting from $X_1 \neq X_2$, X_1 and $X_2 \in \hat{\Omega}' \setminus \Omega$, consider a continuation $\bar{f}$ of an f on $\hat{\Omega}'$, separating X_1, X_2; f has different limits according to $\mathcal{F}_1, \mathcal{F}_2$ and a limit according to $\mathcal{F}_{\Omega}$. It is a contradiction.

We conclude $\mathcal{F}_{\Omega}$ converges in $\hat{\Omega}'$ to $X \in \hat{\Omega}' - \Omega$ and by commuting $\hat{\Omega}$, $\hat{\Omega}'$, we see that the correspondence $x \to X$ is bijective. Now, the identity mapping $x \to x$ of Ω of the space $\hat{\Omega}$, on Ω of the space $\hat{\Omega}'$ has the limit $X \in \hat{\Omega}' - \Omega$ at any $x \in \hat{\Omega} - \Omega$; there is therefore a continuous continuation which defines a homeomorphism between $\hat{\Omega}$ and $\hat{\Omega}'$.

It remains to prove the existence of such an $\hat{\Omega}$. Let Φ_0 be the set of all finite continuous functions on Ω with compact support. Consider the set $\Psi = \Phi \cup \Phi_0$. Let $\bar{R}_{\psi}$ ($\psi \in \Psi$) be a space $[-\infty, \infty]$ corresponding to ψ and let $A = \prod_{\psi \in \Psi} \bar{R}_{\psi}$, which is compact. Let m be the mapping of Ω in this space, defined by $m(x)$ = point of coordinate $\psi(x)$ on every $\bar{R}_{\psi}$. We will prove it defines a homeomorphism.

a) it is obviously continuous

b) it is injective : given x_1, $x_2 \neq x_1$, there exists a $\psi \in \Phi_0$ equal to zero and 1 at these points (use the property that Ω is uniformizable or that its Alexandroff compactification is normal).

c) The inverse mapping m^{-1} from $m(\Omega) \to \Omega$ is continuous: We consider a neighbourhood $\mathcal{V}$ of $x_0 \in \Omega$ in Ω and we will prove that $m(\mathcal{V})$ is a neighbourhood of $m(x_0)$ in $m(\Omega)$. Consider in $\mathcal{V}$, a compact neighbourhood U of x_0, in Φ_0 a function ψ_0, $\neq 0$ at x_0 and whose support is in U, and in $m(\Omega)$, the set E of the points whose coordinate on $\bar{R}_{\psi_0}$ is $\neq 0$. E is open in $m(\Omega)$ (inverse image of an open set) and contains $m(x_0)$. If $m(x) \in E$, its projection on $\bar{R}_{\psi_0}$ is $\psi_0(x) \neq 0$, hence $x \in U$, $m(x) \in m(U)$, $E \subset m(U)$; therefore $m(U)$ is a neighbourhood of $m(x_0)$ in $m(\Omega)$ and $m(\mathcal{V})$ too.

Now Ω is dense in a compact space $\hat{\Omega}$ homeomorphic to $\overline{m(\Omega)}$ (closure in A) (consider on $m(\Omega)$, the uniform structure of the compact space $\overline{m(\Omega)}$, the corresponding structure on Ω and the completion of both spaces).

It will be easy now to prove that this $\hat{\Omega}$ satisfies (ii) and (iii). Consider $f \in \Phi$. The projection of $m(x)$ on $\bar{R}_f$ is $f(x)$; this mapping of $m(\Omega)$ in $\bar{R}_f$ has a continuous continuation according to the projection of $\overline{m(\Omega)}$ in $\bar{R}_f$. Therefore, by homeomorphism, f has a continuous continuation on $\hat{\Omega}$.

Consider finally on $\hat{\Omega} - \Omega$, two points x_1, x_2, $x_1 \neq x_2$; the images on $\overline{m(\Omega)} - m(\Omega)$ have different projections on a suitable $\bar{R}_\psi$, i.e. $\psi(x_1) \neq \psi(x_2)$. Hence $\psi \notin \Phi_0$; if not, the continuous continuation of ψ on the boundary of Ω would be 0. We conclude $\Psi \in \Phi$; (iii) is proved.

2. Other characterisations of this compactification.

Theorem XIII, 2. As a complement of the previous theorem, the uniform structure on Ω induced by the unique one of $\hat{\Omega}$ is characterized.

a) as the coarsest one S making uniformly continuous the functions of Ψ.

b) as the coarsest one S' making uniformly continuous the functions of Φ and compatible with the topology of Ω.

Consequently, $\hat{\Omega}$ is the completion of Ω according to S or S'.

Proof. We will see first the equivalence of (a) and (b). The topology given by S is the coarsest one making continuous the functions of Ψ and this is the topology of Ω. Hence S is a structure among those considered in (b). It is finer than S'. But the structures of (b) are in the family of the structures of (a). Hence S' is finer than S.

Now S is precompact because the functions of Ψ are mappings in a compact space. Therefore, the corresponding completion $\bar{\Omega}$ is compact; let us see it is homeomorphic to $\hat{\Omega}$:

The neighbourhoods of $x \in \hat{\Omega} - \Omega$ intersect Ω according to sets of a Cauchy filter with the uniform structure of $\hat{\Omega}$ (for which the functions of Ψ are uniformly continuous). Therefore with S which is coarser, this filter converges in $\bar{\Omega}$ to some $X \in \bar{\Omega} - \Omega$. This mapping $x \to X$ is injective: consider $x_1 \neq x_2$ on $\hat{\Omega} - \Omega$ and $f \in \Phi$ whose continuous continuation on $\hat{\Omega}$ separates

x_1 and x_2; we must have $X_1 \neq X_2$; if not, f which has a continuous continuation in $\bar{\Omega}$, would have the same limit according to the filters on Ω induced by the neighbourhoods of x_1, x_2 in $\hat{\Omega}$. The mapping is even bijective: the neighbourhoods of $X \in \bar{\Omega} - \Omega$ in $\bar{\Omega}$ intersect Ω according to the sets of a filter whose suitable finer filter $\mathcal{G}$ will converge in $\hat{\Omega}$ to some point x of $\hat{\Omega} - \Omega$ and the image by the considered mapping is the limit of this $\mathcal{G}$ in $\bar{\Omega}$, i.e. X. The identity mapping $x \to x$ of Ω (considered in $\hat{\Omega}$) on Ω (considered in $\bar{\Omega}$) has limits at the points of $\hat{\Omega} - \Omega$, i.e. has a continuous continuation; it is a one-one continuous mapping of $\hat{\Omega}$ (compact) on $\bar{\Omega}$ (Hausdorff), therefore, it is a homeomorphism.

<u>Exercise</u>: Consider only S' and prove directly the corresponding completion is $\hat{\Omega}$.

3. <u>Examples of applications</u>: (See Constantinescu-Cornea [1]).

1) Theorem (XIII, 1) when Φ is empty, means that $\hat{\Omega}$ is the unique compact space, where Ω is dense, and containing only one point outside of Ω. $\hat{\Omega}$ is therefore, the Alexandroff compactification (upto homeomorphism).

2) When Φ contains all continuous real (finite or not) functions, we get the <u>Stone-Čech compactification</u>.

3) If the functions of Φ are all real continuous functions such that for every f, there exists a compact set whose complementary set is the union of domains for which f is constant on every one, we get the <u>compactification of Kerekjarto-Stoïlov</u>, given first for Riemann Surfaces, and commonly used in the theory of a complex variable.

4) With all real continuous BLD functions (see Chap IX, §, footnote 39) on a Riemann surface or even on a $\mathcal{E}$-space, we get the <u>Royden compactification</u>.

5) Let us consider the following subclass of the previous functions: for such a f, there exists a closed set F such that f is harmonic on $\complement F$, and realizes the minimum of the Dirichlet integral for the BLD functions equal to f on F.

So, we get the <u>Kuramochi compactification</u> (originally for Riemann

Surfaces) specially adapted to the study of the BLD functions. On this important question, see the papers of a Symposium (Lecture Notes no. 58, Springer).

6) In the next chapter, we will study with details another and more important compactification, the Martin one.

CHAPTER XIV

Classical Martin Space[45]

The Martin Integral Representation

1. Elementary Introduction. On a Green space Ω, we consider the family with parameter y of functions $x \rightsquigarrow \frac{G(x,y)}{G(x, y_0)} = K(x, y)$ (y_0 fixed in Ω). With the convention that $K(y_0, y_0) = 1$, we may see that $x \rightsquigarrow K(x,y)$ is continuous even when $y = y_0$.

Theorem XIV, 1. The space $\hat{\Omega}$ introduced by the Constantinescu-Cornea theorem (Chap XIII) from the previous family is, upto a homeomorphism, independant of y_0. It is called the Martin space and $\Delta = \hat{\Omega} - \Omega$, the Martin boundary. The topology of $\hat{\Omega}$ will be denoted $\mathcal{C}_m$.

Proof. On any domain $\Omega_1 \subset \bar{\Omega}_1 \subset \Omega$, $\Omega_1 \ni y_0$, $y \rightsquigarrow \frac{G(x,y)}{G(x,y_0)}$ is harmonic when $x \notin \Omega_1$, >0 and is equal to 1 at y_0. When $x \longrightarrow X \in \hat{\Omega} - \Omega$, this function converges therefore to a harmonic > 0, function. Consider

$$\frac{G(x,y)}{G(x,y_0')} = \frac{G(x,y)}{G(x,y_0'')} \cdot \frac{G(x, y_0'')}{G(x, y_0')} \text{ , where } y_0', y_0'' \in \Omega_1, y_0' \neq y_0'' .$$

If $\hat{\Omega}'$, $\hat{\Omega}''$ correspond to the families relative to y_0', y_0'', both right functions converge to finite > 0 limits when $x \longrightarrow X \in \hat{\Omega}'' - \Omega$ in $\hat{\Omega}''$, therefore also the left member. For such different X_1, X_2, if the limits of the left member are different for $y = y_0''$, these limits separate $\hat{\Omega}'' - \Omega$; if not, for a suitable y, the limits of $\frac{G(x,y)}{G(x,y_0'')}$ are different by definition of $\hat{\Omega}''$. We conclude that the limits of the first member for $x \longrightarrow X \in \hat{\Omega}'' - \Omega$ (in $\hat{\Omega}''$) and all y, exist and separate $\hat{\Omega}'' - \Omega$. Therefore, $\hat{\Omega}'$ is homeomorphic to $\hat{\Omega}''$.

Complements i) The limit $K(X,y)$ of $K(x,y)$ as $x \longrightarrow X \in \hat{\Omega} - \Omega$ (denoted also $K_X(y)$) is real continuous of $(X, y) \in \Delta \times \Omega$.

---

(45) Fundamental Paper of R.S. Martin [1]. See a survey in Brelot [12],[24].

That is a consequence of the continuity in X and of the equicontinuity in y.

ii) With a fixed y_o, we may consider the family of $x \rightsquigarrow K(x,y_i)$ for a dense set $\{y_i\}$. The space which corresponds by the Constantinescu-Cornea theorem is still $\hat{\Omega}$, because $y \rightsquigarrow K(x,y)$ converges when x tends to a boundary point of this space. Moreover, we may introduce a countable family of real continuous functions φ_j with compact support, which is dense on the set of such functions (norm = sup $|\ |$). The coarsest uniform structure on Ω which makes uniformly continuous the $K(x, y_i)$ and the φ_j, is the uniform structure whose completion gives $\hat{\Omega}$; and that proves that this structure has a countable base of "entourages". Therefore, $\hat{\Omega}$ is metrizable.

Exercise: Given an explicit metric (as Martin did).

iii) Instead of $G(x, y_o)$, we may introduce $G'(x,y_o) = \int G(x,y)d\rho_{y_o}^{\omega_o}(y)$ ($d\rho_{y_o}^{\omega_o}$ harmonic measure for the regular domain $\omega_o \ni y_o$) continuous, and a corresponding K', whose associated space is still $\hat{\Omega}$ ($K' = K$ for $x \notin \omega_o$).

2. The minimal harmonic functions on a Green space Ω.

Let us take for U and P (Chap XII, § 3), the non-negative harmonic functions and the potentials. Axioms A_1, A_2 are fulfilled. Moreover, U is convex and in the space $\mathcal{E}_U$ of the differences of non-negative harmonic functions, with the natural order, U is the non-negative cone; the minimal harmonic functions are the points of the extreme generatrices of U. On $\mathcal{E}_U$, we may introduce the topology of the local uniform convergence and the set of functions of $\mathcal{E}_U$, defined by $u(y_o) = 1$ is a closed hyperplane; the positive ones form a compact metrizable base B_{y_o} (for example with distance = sup $|u_1 - u_2|$ on a fixed neighbourhood of y_o). The extreme points are exactly the minimal harmonic functions equal to 1 at y_o, and form a G_δ set (extreme points of a compact metrizable set in a vector space) which is not empty.

Let us determine the minimal harmonic functions.

It is obvious that if μ is a positive measure on Δ, $\int K(X,y)d\mu(X)$ is a positive harmonic function on Ω (apply the mean criterion). Conversely,

<u>Lemma XIV, 2.</u> Any harmonic function $u \geqslant 0$ on Ω may be represented by $u(y) = \int K(X, y)\, d\mu(X)$, where μ is a positive Radon measure on Δ.

<u>Proof</u>. If Ω_n is an increasing sequence of relatively compact open sets $\Omega_n \subset \bar{\Omega}_n \subset \Omega$, with $\bigcup \Omega_n = \Omega$, $\hat{R}_u^{\Omega_n}$ is a potential (superharmonic and $\leq \lambda G_{y_0}$ where λ satisfies $\lambda G_{y_0} \geqslant u$ on $\partial\Omega_n$, therefore on Ω_n as well).

Thanks to the Riesz representation,

$$\hat{R}_u^{\Omega_n}(y) = \int G(x,y)\, d\mu_n(x)$$
$$\int K(x,y)\, d\nu_n(x)$$

with $d\nu_n(x) = G(x, y_0)\, d\mu_n(x)$ and $\int d\nu_n = u(y_0)$, support of $\nu_n \subset \partial\Omega_n$. We may extract a subsequence ν_{n_p} converging vaguely to a positive measure μ on $\hat{\Omega}$, with its support on Δ. The continuity of $x \rightsquigarrow K(x,y)$ on $\hat{\Omega}$ gives

(1) $u(y) = \int K(X, y)\, d\mu(X)$.

<u>Theorem XIV, 3</u>. <u>Any minimal positiveharmonic function</u> u <u>is equal to</u> $u(y_0)K(X,y)$ <u>for some</u> $X \in \Delta$.

The corresponding X are all called minimal points of Δ, and their set is denoted Δ_1.

<u>Proof</u>. Let $u > 0$. We use the representation $u(y) = \int K(X,y)\, d\mu(X)$ where μ is a strictly positive measure. On Δ, there exists a point X_0 such that any open neighbourhood $\mathcal{V}$ of X_0 in $\hat{\Omega}$, has a μ-measure $\neq 0$. Hence,

$$\int_{\mathcal{V}} K(X, y)\, d\mu(X) = \lambda u(y) \quad \text{with } \lambda = \frac{\mu(\mathcal{V})}{u(y_0)}$$

where $\mathcal{V}$ is any neighbourhood of X_0. Hence

$$\frac{u(y)}{K(X_0,y)u(y_0)} = \frac{\int_{\mathcal{V}} \frac{K(X,y)}{K(X_0,y)}\, d\mu(X)}{\int_{\mathcal{V}} d\mu} .$$

Let y be fixed. Given $\varepsilon > 0$, $1 - \varepsilon < \frac{K(X,y)}{K(X_0,y)} < 1 + \varepsilon$ for $X \in \mathcal{V}$ with a suitable $\mathcal{V}$. The same holds for the second member.

Hence, the first member is equal to 1.

Remark: The previous argument proves that if $\int K(X, y)\, d\mu(X)$ is minimal, it is proportional to $K(X_0, y)$ where X_0 is a point of the closed support of μ on Δ.

Correspondance $X \rightsquigarrow K_X$ between Δ and the set B'_{y_0} of the K_X on B_{y_0}.

It is a homeomorphism because this mapping $X \rightsquigarrow K_X$ of Δ (compact) on B'_{y_0} (Hausdorff space) is continuous and one-one. The image of Δ_1 is the set of the extreme points of B_{y_0}.

3. The Martin representation.

Recall the fundamental results of Choquet:-

In a locally convex Hausdorff topological vector space E, let us consider a convex cone C with a compact base B (intersection of C with a closed hyperplane).

α) If B is metrizable, any $X \in B$ is the barycentre of a measure $\mu \geqslant 0$ ($\|\mu\| = 1$) supported by the set $\mathcal{E}$ of the extreme points of B(i.e. $\mu(\complement\mathcal{E}) = 0$). That means that for every continuous linear form l on E,

$$l(X) = \int l(x)\, d\mu(x).$$

β) If for the own order of C, C is a lattice, such a representation is unique.

In our case, with the space $\mathcal{E}_U$, the cone U, and base B_{y_0}, we know for B_{y_0}, the compactness and metrizability and for C the property of lattice (Chap VI), which is possible to prove without the Riesz representation.

Therefore, for any harmonic function $u \geqslant 0$,

(2) $\quad l(u) = \int l(v)\, d\mu(v) \quad \{v \in B_{y_0}, \mu$ a + ve Radon measure on $B_{y_0}, \mu(\complement\mathcal{E}) = 0, \mu$, measure also on $B'_{y_0}\}$.

(3) $\quad u(y) = \int v(y)\, d\mu(v),\ v \in B_{y_0}$, same μ which is independant of y.

As Δ is homeomorphic to B'_{y_0}, μ may be considered as a measure on Δ.

Hence

Theorem XIV, 4. For any harmonic function $u \geqslant 0$ on Ω

$$(4) \qquad u(y) = \int K_X(y)\, d\mu(X),\ X \in \Delta,\ \mu \geq 0$$

<u>on Δ supported by Δ_1 and is unique.</u>

The uniqueness in (2) implies the same in (3) or (4) because (3) for all y implies (2) for all 1. If not, u would not be the barycentre of μ; this barycentre is a point of B_{y_0}, i.e. a harmonic function different from u and at a suitable point y, $u(y) \neq V(y) = \int v(y)\, d\mu(v)$.

4. <u>Examples:- Case of a ball in</u> $\mathbb{R}^n$.

<u>Theorem XIV, 5. In the case of a ball</u> (<u>centre</u> y_0, <u>radius</u> R), $\dfrac{G(x,y)}{G(x,y_0)}$ <u>has a limit when</u> $x \to X$, <u>point of the ordinary Euclidean boundary, and this limit is</u>

$$(5) \qquad K_X(y) = R^{n-2}\, \frac{R^2 - |y-y_0|^2}{|X-y|^n}.$$

<u>Hence the euclidean closure of this ball</u> Ω <u>and the boundary are homeomorphic to</u> $\hat{\Omega}$ <u>and</u> Δ; $K(X, y)$ <u>is the previous Poisson kernel</u>. <u>Moreover</u> $\Delta_1 = \Delta$ (<u>because all boundary points must be of the same type and at least one is minimal</u>).

<u>Proof.</u> It is easier to work first with the half-space.

In $\mathbb{R}^3$, for the half-space, $G(x,y) = \dfrac{1}{|x-y|} - \dfrac{1}{|x_1-y|}$ where x_1 is the symmetric point of x relative to the boundary plane $\mathcal{H}$.

With a fixed y_0,

$$\frac{G(x,y)}{G(x,y_0)} = \frac{|x_1-y| - |x-y|}{|x_1-y_0| - |x-y_0|} \cdot \frac{|x-y_0| \cdot |x_1-y_0|}{|x-y| \cdot |x_1-y|}.$$

The first factor is equal to

$$\frac{|x_1-y|^2 - |x-y|^2}{|x_1-y_0|^2 - |x-y_0|^2} \cdot \frac{|x_1-y_0| + |x-y_0|}{|x_1-y| + |x-y|}.$$

Now, thanks to an elementary theorem of Geometry $|x_1-y|^2 - |x-y|^2 = 4\,\delta_x \delta_y$ where δ_x stands for the distance of x from $\mathcal{H}$.

We conclude easily that as

$$x \to X \in \mathcal{H}, \ \frac{G(x,y)}{G(x,y_0)} \to \frac{\delta_y}{\delta_{y_0}} \cdot \frac{|X-y_0|^3}{|X-y|^3} .$$

In $\mathbb{R}^2$, we use $\log x \sim x-1$ as $x \to 1$ and a similar calculation gives the limit $\frac{\delta_y}{\delta_{y_0}} \cdot \frac{|X-y_0|^2}{|X-y|^2}$.

In $\mathbb{R}^n$ $(n > 3)$, we use the identity $a^n - b^n = (a-b)(a^{n-1}+a^{n-2}b+\dots+b^{n-1})$ in order to treat $\frac{1}{|x_1-y|^{n-2}} - \frac{1}{|x-y|^{n-2}}$; by coming back to $\frac{1}{|x_1-y|^{n-2}} - \frac{1}{|x-y|^{n-2}}$ and using the previous calculations, the general result is

$$(6) \qquad \frac{G(x,y)}{G(x,y_0)} \to \frac{\delta_y}{\delta_{y_0}} \frac{|X-y_0|^n}{|X-y|^n} \quad (x \to X \in \mathcal{H}).$$

The same argument shows that when x tends to the Alexandroff point of $\mathbb{R}^n$, $\frac{G(x,y)}{G(x,y_0)} \to \frac{\delta_y}{\delta_{y_0}}$; therefore (6) is valid for all boundary points with an ordinary obvious convention. That gives the "Poisson Kernel" of the half-sapce. With a Kelvin transformation (simple inversion in $\mathbb{R}^2$) we conclude that a limit for $\frac{G(x,y)}{G(x,y_0)}$ exists at every boundary point in the case of a ball. Hence the identification of $\hat{\Omega}$ with the Euclidean closure.

In order to get the expression of the limit, it is sufficient and simpler to use the previous limit property for the Alexandroff point.

We have to take the Kelvin transformation of $\frac{\delta_y}{\delta_{y_0}}$, with the inversion corresponding to the ball of radius δ_{y_0}, centre at the point X_0, symmetric point of y_0 relative to the hyperplane $\mathcal{H}$ and tangent at Z_0. The image of the half space is a ball of centre $Y_0 = \frac{X_0 + Z_0}{2}$ and radius $R = \frac{\delta_{y_0}}{2}$. Let us take y'_0 for origin and denote by y_i, y'_i the coordinates of y and its image y', the last axis being normal to $\mathcal{H}$ (with $y_n > 0$). Then

$$\delta_y + 2R = \delta^2_{y_o} \frac{(y'_n + R)}{|X_0 - y'|^2}$$

$$\frac{\delta_y}{\delta_{y_0}} = \delta_{y_0} \frac{y'_n + R}{|X_0 - y'|^2} - 1$$

$$= \frac{2 R y'_n + 2 R^2 - |X_0 - y'|^2}{|X_0 - y'|^2} .$$

As
$$|y' - X_0|^2 = |y' - y'_0|^2 + |y'_0 - X_0|^2 + 2(\overrightarrow{y' - y'_0}) \cdot (\overrightarrow{y'_0 - X_0})$$

$$= |y' - y'_0|^2 + R^2 + 2 R y'_n$$

$$\frac{\delta_y}{\delta_{y_0}} = \frac{R^2 - |y' - y'_0|^2}{|X_0 - y'|^2} .$$

In $\mathbb{R}^n$, the Kelvin transformed is

$$\frac{R^{n-2}}{|X_0 - y'|^{n-2}} \cdot \frac{R^2 - |y' - y'_0|^2}{|X_0 - y'|^2} = R^{n-2} \frac{R^2 - |y' - y'_0|^2}{|X_0 - y'|^n} .$$

We now conclude to the general form (5) of the wanted limit.

Corollary. An immediate consequence is the Lebesgue-Stieltjes representation of the positive harmonic functions in the ball of centre y_0, radius R

$$(7) \qquad u(y) = \int R^{n-2} \frac{R^2 - |y - y_0|^2}{|X - y|^n} \, d\mu(X)$$

(unique positive measure μ on $\partial\Omega$).

Exercises: 1) The existence of such a μ may be deduced from the representation in a concentric smaller ball by means of the Poisson integral, thanks to a limit process.

2) From the existence of such a representation, we deduce that the minimal positive harmonic functions are exactly all functions proportional to the previous Poisson kernel.

Remark. In the case of a ball, the property of the limit of $\frac{G(x,y)}{G(x,y_0)}$ (y_0 center) when x tends to a boundary point contains the existence of a normal derivative of $x \rightsquigarrow G(x,y)$ at X, (because $G(x, y_0)$ is then equivalent to $\frac{R - |x-y_0|}{R^{n-1}}$)

and this derivative $\dfrac{\partial G(x,y)}{\partial n_X}$ is equal to $\dfrac{1}{R}\;\dfrac{R^2-|y-y_0|^2}{|X-y|^n}$. Recall that for a harmonic function on a side of a smooth surface (twice continuously differentiable) and vanishing on this set, the gradient has a finite limit on this set and in our case of $x \rightsquigarrow G(x,y)$ in a ball, it tends to $\dfrac{\partial G}{\partial n_X}$, which we have studied independantly.

We may underline that the classical solution of the Dirichlet problem for the ball i.e., the Poisson integral

$$\int K_X(y)\; f(X)\; d\nu(X) \quad \text{or} \quad \int R^{n-2}\;\frac{R^2-|y-y_0|^2}{|X-y_0|^n}\; f(X)\; d\nu(X)$$

(ν distribution of mass 1 on the sphere) may be written $R^{n-1}\displaystyle\int \frac{\partial G}{\partial n_X}(y) f(X) d\nu(X)$

and the general representation (1-7) as well $\displaystyle\int \frac{\partial G}{\partial n_X}\,(y)\; dm(X)$.

5. Another fundamental example.

Theorem XIV, 6. In a Green space Ω, consider a polar point x_0 and $\Omega_1 = \Omega \setminus \{x_0\}$. The minimal harmonic functions on Ω_1 are the minimal functions of Ω and the functions proportional to $G^{\Omega}_{x_0}$.

Proof. Any minimal function w on Ω has a restriction on Ω_1, which is minimal for Ω_1, because any harmonic minorant $\geqslant 0$ on Ω_1 has a harmonic continuation which minorizes w and hence must be proportional.

Let us consider any u harmonic $\geqslant 0$ on Ω_1; it has a superharmonic continuation $\hat{u}$ on Ω whose Riesz representation is

$$\hat{u} = \lambda G_{x_0} + v \geqslant 0 \quad (v \text{ harmonic})$$

Hence $\lambda \geqslant 0$, $-v \leq \lambda G_{x_0}$, $v \geqslant 0$.

If $u \leq G_{x_0}$, we get $v \leq G_{x_0}$ on Ω_1, then on Ω. Hence $v = 0$, $u = \lambda G_{x_0}$. Therefore, G_{x_0} is minimal on Ω_1. Suppose now u is minimal on Ω_1; if u is bounded close to x_0, λ must be zero, $\hat{u}$ is harmonic and must be minimal on Ω, because any harmonic minorant must be proportional on Ω_1, therefore on Ω. If u is not bounded, the minorant v must be proportional to u, therefore equal to zero and $u = \lambda G_{x_0}$.

Exercise: If x_0 is not polar in the Green space Ω, the minimal harmonic functions on $\Omega_1 = \Omega \setminus \{x_0\}$ are the functions $u - \lambda G_{x_0}$ (u minimal on Ω, λ such that $u(x_0) - \lambda G_{x_0}(x_0) = 0$) and the functions proportional to G_{x_0}.

More difficult proof using the behaviour of the functions harmonic in the neighbourhood of the point at infinity of $\mathbb{R}^n$ ($n \geqslant 3$), this point excluded.

6. Remarks. The study of the ball has to be extended to domains with more or less smooth boundaries, but has not been deeply developed till now.

We will chiefly underline that in $\mathbb{R}^2$ or on a Riemann surface any conformal mapping preserves the Green function, therefore also $K(x,y)$; as a simply connected domain ω whose boundary in $\overline{\mathbb{R}}^2$ contains more than a point, may be conformally mapped onto an open disc, the Martin space of ω is homeomorphic to the Euclidean closure of the disc and the Martin boundary is the set of the prime ends of caratheodary.

In general, for any Euclidean Greenian domain, the Martin topology cannot be compared with the Euclidean one, as examples show it (Martin [1], Brelot [12]). Recall the case of the domain between 2 tangent spheres of $\mathbb{R}^3$; then $K(x,y)$ has many different limit functions when x tends to the tangent point, i.e. there are many Martin points corresponding to this one (Bouligand [1]). Conversely, it may happen for suitable domains that $K(x,y)$ has the same limit function for 2 sequences x_n tending to different euclidean boundary points.

This explains the insufficiency of the ordinary boundary, known for a long time and the success of the new one.

7. Another Characterisation of the Martin Space. Short Indications.

Consider on the Green space Ω, the cone S^+ of the non-negative superharmonic functions and the vector space S of the differences of such functions. More precisely (see Chap. XI). We consider the set of all pairs (u,v) where $u,v \in S^+$. We introduce an equivalence relation on this set as follows. We say

$(u_1,v_1) \sim (u_2,v_2)$ if $u_1+v_2 = u_2+v_1$ i.e. if $u_1 - v_1 = u_2 - v_2$ q.e. (both sides being defined q.e.). The equivalence class corresponding to (u,v) is denoted $[u,v]$. S will be the set of these equivalence classes and S^+ is in one-one correspondance with the set of the $[u,0]$. We said already that S^+ is a (complete, lattice for its proper order (specific order) (possible to prove without the Riesz representation theorem) and that there exists a topology on S making S^+ locally compact. Here, it is sufficient to introduce a countable base of regular domains ω_i and a countable dense set of x_j. $[u,v] \to \left|\int u d\rho_{x_j}^{\omega_i} - \int v d\rho_{x_j}^{\omega_i}\right|$ is a semi-norm and these seminorms define a topology (independant of the choice of the ω_i, x_j). Hence, there exists for S^+ a compact metrizable base B, for instance with the condition that $\int u d\rho_{y_0}^{\omega_0} = 1$, $u \in S^+$ (ω_0 regular, y_0 fixed) (See Brelot [19], Mrs. Hervé [1]). We have already mentioned - even in an axiomatic frame that the Choquet theorem may be used and leads the representation of any superharmonic function $u \geqslant 0$ as

(8) $$u(y) = \int v(y)\, d\mu(v), \; v \in B$$

where μ is a + ve measure on B, supported by the set of the extreme points of B. From that, one may deduce that the extreme elements of B are exactly the functions $y \leadsto G(x,y)$ upto a factor i.e. $\lambda_x G(x,y)$ with $\lambda_x = \dfrac{1}{\int G(x,y) d\rho_{y_0}^{\omega_0}(y)}$ and the minimal harmonic functions equal to 1 at y_0.

Theorem XIV,7. The set E of the potentials belonging to B with point support (i.e. the $\lambda_x G(x,y)$) and its closure $\bar{E}$ in B are homeomorphic to Ω and $\hat{\Omega}$ respectively.

Proof. For the first point (see Brelot [18], Gowrisankaran [1]). Then we see that $\bar{E}$ is homeomorphic to a compact $\check{\Omega}$ where Ω is dense. For a sequence of functions belonging to B and which are harmonic in an open set ω, the convergence on B implies that the sequence converges at every point of ω. Hence $\check{\Omega}$ satisfies all conditions defining $\hat{\Omega}$ (essentially, the existence of a limit for

$\frac{G(x,y)}{G(x,y_0)}$ when x tends to a point of $\check{\Omega} - \Omega$).

Recall that in (8) μ may be decomposed into μ_1 supported by the set of the extreme potentials and μ_2, supported by the set of the minimal harmonic functions. Hence, thanks to the homeomorphism of $\mathcal{E}$ and Ω and that of $\bar{\mathcal{E}} - \mathcal{E}$ and Δ we get:

(9) $$u(y) = \int G(x,y)\, \lambda_x \, d\mu_1(x) + \int K_X(y)\, d\mu_2(X)$$

(decomposition of Riesz-Martin).

The second part is the greatest harmonic minorant whose representation was given in (1).

CHAPTER XV

Classical Martin Space and Minimal Thinness

1. Minimal thinness. Minimal fine topology on $\Omega \cup \Delta_1$.

We will apply Chapter XII. On a Green space Ω, we consider the convex cone U of the non-negative harmonic functions, the convex cone P of the potentials, the cone Φ of the non-negative harmonic functions. Every class of proportionality of minimal harmonic functions correspond to an index point $X \in \Delta_1$; the general abstract minimal boundary $\mathfrak{m}$ of Chap. XII. § 4 becomes here Δ_1. The thinness of $e \subset \Omega$ at X will be therefore characterized by the condition $R^e_{K_X} \neq K_X$ or by the existence of a potential majorizing K_X on e (or only q.e on e and whose measure is supported by B_e). As in general the union of two thin sets is thin and Ω is unthin, the complements of such thin sets e form a filter $\mathcal{G}_X$.

Remark. When e is thin at X, $\hat{R}^{e \cap \complement K}_{K_X}$ is a potential (for any compact $K \subset \Omega$) and tends to zero according to the increasing directed set of these K. In particular if δ_n is a neighbourhood of X on $\hat{\Omega}$ with $\bigcap \delta_n = X$, $R^{e \setminus \delta_n}_{K_X} \longrightarrow 0$.

Exercise. Any F_σ set e in Ω is thin at $X \in \Delta_1$ when any closed subset is thin (indicated by Toda for an open e).

Example. In the case of a ball Ω the set $\{y : K(X,y) < \lambda\}$ is the complement of another contained ball (tangent at X) and is (minimal) thin at X, but not thin in the ordinary meaning in $\mathbb{R}^n$.

In the general frame of Chap XII § 4 we have considered the possible continuations of the fine topology on Ω by a topology on $\Omega \cup \Delta_1$ for which the neighbourhoods of any $X \in \Delta_1$ cut Ω according to the sets of $\mathcal{G}_X$. We have seen there was a finer one and among those making Ω open, a coarsest one. We shall precise that in our particular case:

Lemma XV, 1. If δ is an open neighbourhood of $X \in \Delta_1$ on $\hat{\Omega}$, $\Omega \setminus \delta$ is thin at X (hence $\delta \cap \Omega$ is unthin at X).

<u>Proof.</u> If not $K_X \equiv R_{K_X}^{\Omega \setminus \delta} \equiv \hat{R}_{K_X}^{\Omega \setminus \delta}$. But if Ω_n is an increasing sequence of relatively compact open sets such that $\Omega_n \subset \bar{\Omega}_n \subset \Omega, \bigcup \Omega_n = \Omega$, then $\hat{R}_{K_X}^{\Omega_n \setminus \delta} \longrightarrow \hat{R}_{K_X}^{\Omega \setminus \delta}$. As $\hat{R}_{K_X}^{\Omega_n \setminus \delta}$ is a potential V_n, it may be represented by

$$\int G(x,y)\, d\mu_n(x) \text{ or } \int K(x,y)\, d\nu_n(x)$$

with $\int d\nu_n = 1$. A suitable subsequence ν_{n_p} converges vaguely to ν on $\hat{\Omega}$. Hence

$$(1) \qquad K_X(y) = \int K(x,y)\, d\nu(x),\ y \in \Omega,\ x \in \hat{\Omega}$$

$\int d\nu = 1$ (Use $y = y_0$). As $\nu_n(\delta \cap \Omega_n) = 0$ (harmonicity of V_n on $\delta \cap \Omega_n$), ν_n is supported by $\complement_{\hat{\Omega}} \delta$, hence also ν (46). The harmonicity of the integral (1) in Ω implies that $\nu(\Omega) = 0$ and that ν is supported by $\Delta \cap \complement_{\hat{\Omega}} \delta$. As this integral is minimal, the remark (Chap XIV, end § 2) shows that it must be equal to a $K_{X_0}(y)$ with $X_0 \in \complement_{\hat{\Omega}} \delta$. Contradiction to the unicity of such a X and to (1) with $X \in \delta$.

<u>Lemma XV, 2.</u> For every $X_0 \in \Delta_1$, there exists on $\hat{\Omega}$ an open set α containing $\Delta \setminus X_0$ such that $\alpha \cap \Omega$ is thin at X_0 (and is of course unthin at every $Y \in \Delta_1$, $Y \neq X_0$).

<u>Proof.</u> We cover $\hat{\Omega} \setminus X_0$ by a countable union of open sets α_n whose closure does not contain X_0. $\alpha_n \cap \Omega$ is unthin at any point of $\Delta_1 \cap \alpha_n$ and thin at X_0. $\hat{R}_{K_{X_0}}^{\alpha_n \cap \Omega}$ is a potential; for a suitable $\omega_n \subset \bar{\omega}_n \subset \Omega$, $\alpha'_n = \alpha_n \cap \Omega \cap \omega_n$ gives

$$R_{K_{X_0}}^{\alpha'_n \cap \Omega}(y_0) < 2^{-n}\varepsilon,\quad R_{K_{X_0}}^{\bigcup \alpha'_n \cap \Omega}(y_0) < \varepsilon \ (\varepsilon \text{ chosen} < 1).$$

$\bigcup \alpha'_n$ satisfies the wanted conditions.

---

(46) We may finish in the following way: the barycenter of ν on B must be K_X as a consequence of (1). As K_X is extreme on B, ν is the mass 1 at $X \in \delta$. Contradiction.

Theorem XV, 3. (Brelot [33]) There exists on $\Omega \cup \Delta_1$, a unique topology such that

(i) it induces on Ω the fine topology and

(ii) the corresponding neighbourhood of any $X \in \Delta_1$ intersect Ω according to the sets of $\mathcal{F}_X$. It is Hausdorff, makes Ω open and induces on Δ_1 the discrete topology.

Proof. If ω is an open relatively compact neighbourhood of $x \in \Omega$, $\complement \omega$ is unthin at any $X \in \Delta_1$. There exists in a topology satisfying (i) and (ii) an open set α such that $\alpha \cap \Omega = \omega$; if $\alpha \cap \Delta_1 \neq \emptyset$, consider $X \in \alpha \cap \Delta_1$; $\Omega \setminus \alpha \subset \complement_{\Omega} \omega$ must be thin at X. Contradiction. Hence $\alpha = \omega$. Ω is open in the considered topology.

Let us see now that the coarsest topology satisfying (i), (ii) (equivalently with the condition that Ω remains open) is identical to the finest one. We will use Chap XII, § 4 and theorem XII, 1 by taking the points of Δ_1 for the $\{i\}$ and the fine open sets of Ω, which are also elements of $\mathcal{F}_X$ as elements of $\mathcal{B}_1$. The existence (lemma XV, 2) of a fine closed set, thin at X, unthin at every other point of Δ_1 shows that $\{X\}$ is a neighbourhood of X in the considered coarsest topology. Hence the wanted identity with the finest one.

We will call the topology of the above theorem as the (minimal) fine topology on $\Omega \cup \Delta_1$.

2. Interpretation of the minimal fine topology on $\Omega \cup \Delta_1$.

Lemma XV, 4. If $E \subset \Omega$ is thin at $X \in \Delta_1$, there exists an open set $\omega \supset E$, which is also thin at X.

Proof. We know that $R^e_{K_X}(y) = \inf_{\omega} R^{\omega}_{K_X}(y)$ (ω open $\supset e$) (See theorem VIII, 12). As $R^e_{K_X}(y_1) < K_X(y_1)$ for a suitable y_1, we conclude that $R^{\omega}_{K_X}(y_1) < K_X(y_1)$ for a suitable ω. Hence ω is thin at X.

Lemma XV, 5. Any inequality $\int G(x,y)\, d\mu_1(y) \leq \int G(x,y)\, d\mu_2(y)$ for every x implies $\int v\, d\mu_1 \leq \int v\, d\mu_2$ for every superharmonic $v \geq 0$.

Proof. If α_n is an increasing sequence of compact sets such that $\bigcup \alpha_n = \Omega$, $\hat{R}^{\alpha_n}_v(y)$ is a potential $\int G(x,y)\, d\nu_n(x)$ and

$$\int \hat{R}_v^{\alpha_n} \, d\mu_1 = \int \left(\int G(x,y) \, d\mu_1 \right) d\nu_n .$$

The same is true with μ_2.

The hypothesis implies

$\int \hat{R}_v^{\alpha_n} \, d\mu_1 \leq \int \hat{R}_v^{\alpha_n} \, d\mu_2$ and making $n \to \infty$, we get the wanted result.

<u>Theorem XV, 6.</u> [Naïm [1]]. <u>The thinness of</u> $E \subset \Omega$ <u>at</u> $X \in \Delta_1$ <u>is equivalent to the existence of a measure</u> $\mu \geqslant 0$ <u>such that, with the Martin topology</u> $\mathcal{T}_{\mathcal{M}}$,

$$\int K_X(y) \, d\mu(y) < \liminf_{\substack{x \in E \\ x \to X}}{}_{\mathcal{T}_{\mathcal{M}}} \int K(x,y) \, d\mu(y). \tag{2}$$

<u>Proof</u>. We know that if X is not on the $\mathcal{T}_{\mathcal{M}}$-closure of E, E is thin and the previous condition is satisfied (for $\mu = 0$), because the right member means

$\sup_{\delta} \inf_{x \in E \cap \delta} \int K(x,y) d\mu(y)$ (δ, a $\mathcal{T}_{\mathcal{M}}$-neighbourhood) which is $+\infty$ (the inf on a void set is $+\infty$).

Suppose E is thin and thanks to lemma XV,4, even open. There exists $Z \in \Omega$ such that $K_X(Z) > \hat{R}^E_{K_X}(Z) = \int K_X(y) \, db^E_{\varepsilon_Z}(y)$ (See Chap. VI § 12). We shall see this measure $b^E_{\varepsilon_Z}$ satisfies (2). Consider

$V_Z(x) = \int K(x,y) \, d\, b^E_{\varepsilon_Z}(y)$ where $y \rightsquigarrow K(x,y)$ is superharmonic > 0. On E,

$$V_Z(x) = K(x,Z) \xrightarrow[x \to X]{\mathcal{T}_{\mathcal{M}}} K(X,Z) > \int K_X(y) d\, b^E_{\varepsilon_Z}(y)$$

which is more precise than (2).

Conversely suppose X is one the $\mathcal{T}_{\mathcal{M}}$ closure of E and μ satisfies (2). We introduce γ strictly between both members and a neighbourhood δ of X in $\hat{\Omega}$ ($\delta \not\ni y_0$) such that

$$\int K(x,y) \, d\mu(y) \geqslant \gamma \quad \text{on } \delta \cap E .$$

Then $\int G(x,y)\, d\mu(y) \geq \gamma\, G(x,y_0)$ on $\delta \cap E$.

For every x, $\int G(x,y)\, d\mu(y) \geq \gamma\, \hat{R}_{G_{y_0}}^{\delta \cap E}(x)$

$$= \gamma\, \hat{R}_{G_x}^{\delta \cap E}(y_0)$$

$$= \gamma \int G(x,y)\, d\, b_{\epsilon_{y_0}}^{\delta \cap E}(y) .$$

Hence

$$K_X(y_0) = 1 > \frac{1}{\gamma} \int K_X(y)\, d\mu(y) \geq \hat{R}_{K_X}^{\delta \cap E}(y_0).$$

We conclude that $E \cap \delta$ is thin at X and E too.

Corollary. For any measure $\mu \geq 0$ on Ω and $X \in \Delta_1$

$$\liminf_{\substack{x \in \Omega \\ x \to X}} {}_{\mathcal{C}_{\mathcal{M}}} \int K(x,y)\, d\mu(y) = \int K_X(y)\, d\mu(y).$$

Theorem XV, 7. On the Martin space $\hat{\Omega}$, let us consider the cone Φ of functions $\int K'(x,y)\, d\mu(y)$, where μ is a positive measure on Ω and the cone Φ_1 of the restrictions on $\Omega \cup \Delta_1$.

i) The fine topology corresponding to Φ_1 on $\Omega \cup \Delta_1$ with the Martin topology $\mathcal{C}_{\mathcal{M}}$ (which is also induced by the fine topology corresponding to Φ on $\hat{\Omega}$) is the minimal fine topology on $\Omega \cup \Delta_1$.

ii) Φ_1 is the set made of $+\infty$ and of the continuation on $\Omega \cup \Delta_1$ by lower-semi-continuity according to $\mathcal{C}_{\mathcal{M}}$ of the functions $\frac{v}{G'(x,y_0)}$, v a potential on Ω.

Proof. ii) is obvious thanks to the last corollary.

The thinness corresponding to Φ_1 at points of Ω is obviously the classical one. For $E \subset \Omega$, it is at any $X \in \Delta_1$, the minimal thinness (previous theorem). According to theorem XV, 3, the fine topology on $\Omega \cup \Delta_1$ corresponding to Φ_1 and $\mathcal{C}_{\mathcal{M}}$ is the minimal fine topology.

Remark. With G_{y_0} instead of $G'(x, y_0)$ and $\frac{v}{G_{y_0}}$ defined at y_0 as equal to lim inf at y_0 on $\Omega \setminus \{y_0\}$, same theorem.

This is a consequence of the property (see Chap IX) that $\frac{v}{G_{y_0}}$ on $\Omega \setminus \{y_0\}$ has a fine limit at y_0 which is equal to the lim inf.

Exercise. Without using theorem XV, 3, prove that the topology corresponding to Φ_1 is the coarsest one on $\Omega \cup \Delta_1$ inducing the classical fine topology on Ω and giving for any $X \in \Delta_1$ neighbourhoods which intersect Ω according to the sets of $\mathcal{J}_X$.

Theorem XV, 8. The family Φ' of the continuations on $\Omega \cup \Delta_1$ by lower-semi-continuity according to $\mathcal{C}_m$ of the functions $\frac{v}{G'(x,y_0)}$ on Ω, v hyperharmonic ≥ 0, **gives on $\Omega \cup \Delta_1$, the same corresponding fine topology as Φ_1, i.e. the minimal fine topology. This is a consequence of the following property**: (**Naïm** [1] **theorem** 7 - 16).

For any superharmonic $v \geq 0$, $\frac{v}{G_{y_0}}$ **on Ω has at any** $X \in \Delta_1$, **a minimal fine limit** (**i.e. according to** $\mathcal{J}_X$) **equal to the** lim $\inf_{\mathcal{C}_m}$.

Proof. Suppose this lim inf, finite, is denoted Λ; we will prove the thinness at X of the set E_ε defined by $\left\{\frac{v(x)}{G(x,y_0)} > \Lambda + \varepsilon,\ \varepsilon > 0,\ x \neq y_0\right\}$. Then, $\forall x \in E$

$$v(x) \geq (\Lambda+\varepsilon)\, \hat{R}^{E_\varepsilon}_{G_{y_0}}(x) = (\Lambda+\varepsilon)\, \hat{R}^{E_\varepsilon}_{G_x}(y_0) = (\Lambda+\varepsilon) \int G_x(y)\, d\, b^{E_\varepsilon}_{\varepsilon_{y_0}} .$$

Hence
$$\frac{v(x)}{(\Lambda+\varepsilon) G_{y_0}} \geq \int K(x,y)\, d\, b^{E_\varepsilon}_{\varepsilon_{y_0}}(y),\ (x \neq y_0).$$

Then
$$\int K_X(y)\, d\, b^{E_\varepsilon}_{\varepsilon_{y_0}}(y) = \liminf_{\substack{x \in \Omega \\ x \to X}} {}_{\mathcal{C}_m} \int K(x,y)\, d\, b^{E_\varepsilon}_{\varepsilon_{y_0}}(y)$$
$$\leq \frac{1}{(\Lambda+\varepsilon)} \liminf_{\substack{x \in \Omega \\ x \to X}} {}_{\mathcal{C}_m} \frac{v}{G_{y_0}} < 1$$

and $\hat{R}^{E_\varepsilon}_{K_X}(y_0) < 1 = K_X(y_0)$.

This proves the wanted thinness.

Remark 1. Another form of the previous development may be given by introducing with L. Naïm [1], a suitable continuation on Δ of $\dfrac{G(x,y)}{G(x,y_0)G(y,y_0)}$. That gives on $\hat{\Omega}$, a kernel Ⓗ(x,y) which is symmetric; the associated potentials of measures ≥ 0, given on $\Omega \setminus \{y_0\}$ (continued by l.s.c at y_0) form a cone whose corresponding fine topology is exactly the minimal fine topology on $\Omega \cup \Delta_1$. This kernel Ⓗ is also useful for the study of the BLD functions (Doob [6]).

Important Remark 2 (Naim [1] theorems 8' - 17). The last result of the previous theorem may be compared to the behaviour of $\frac{v}{K_X}$ $(X \in \Delta_1)$ at X. According to theorem XII, 6, there is a minimal fine limit at X, equal to $\inf \frac{v}{K_X}$; now, it is also obviously equal to $\liminf_{\mathcal{C}_m} \frac{v}{K_X}$ at X. We may add it is also $\mu_v(\{X\})$. This is a consequence of the equality $\mu_v(\{X\}) = \inf_{\Omega} \frac{v}{K_X}$. It is obvious when v is a potential, because it is impossible that $v \geq \varepsilon K_X (\varepsilon > 0)$. Easy when v is harmonic, which may be deduced from the case where $\mu_v(\{X\}) = 0$. Then, we cannot have $v \geq \varepsilon K_X$, which would imply $\mu_v \succ \varepsilon \mu_{K_X}$. As $\inf_{\Omega} \frac{v}{K_X}$ is additive, because equal to a fine limit, the general case follows.

3. Strong Thinness and Unthinness.

Theorem XV, 9. On $\Omega \cup \Delta_1$, with the topology $\mathcal{C}_m$ and the cone Φ_1 of functions, the corresponding thinness and unthinness are always strong.

Proof. That is known on Ω. For a set $E \subset \Omega$, thin at $X \in \Delta_1$, we prove the strong thinness by the existence of a measure μ_0 on Ω such that

$$\int K'_X \, d\mu_0 < +\infty, \int K'(x,y) \, d\mu(y) \underset{\mathcal{C}_m}{\longrightarrow} +\infty \quad (x \in E \; x \to X).$$

We start from a μ satisfying (2); with a suitable decreasing sequence σ_n of open neighbourhoods of X on $\hat{\Omega}$, and the restriction μ_n of this μ on $\sigma_n \cap \Omega$, we see that $\sum \mu_n$ solves the question (see detailed proof Brelot [33]).

As for $\Delta_1 - \{X\}$, it is obviously thin (because the induced topology on Δ_1 is discrete). We may see it is strongly thin by using the set E of lemma 2 and a corresponding measure μ_0 given by the previous statement. Now,

$\int K'(x,y)\, d\mu_0(y)$ which tends to $+\infty$ (according $\mathcal{C}_{\mathcal{M}}$) at X on E, will tend to $+\infty$ on $\Delta_1 - \{X\}$, thanks to the corollary of theorem XV, 6.

As for unthinness, it is sufficient to see that $E \subset \Omega$ is strongly unthin when unthin at X. Denote R^E_φ, the reduced functions relative to Φ_1 on $\Omega \cup \Delta_1$. Any function V of Φ_1 majorizing 1 on $E \setminus \delta$ (δ open neighbourhood of X on $\hat{\Omega}$, $y_0 \notin \bar{\delta}$) is on Ω at least equal to $R^{E\setminus\delta}_{G'_{y_0}} / G'_{y_0}$ which is on $\delta \cap \Omega$ equal to $\hat{R}^{E\setminus\delta}_{G_x}(y_0) / G_x(y_0)$ or to $\int K(x,y)\, d\, b^{E\setminus\delta}_{\varepsilon_{y_0}}(y)$. Now $V(X)$ is $\geq$ $\liminf_{\mathcal{C}_{\mathcal{M}}}$ of this expression, i.e. $\int K(X,y)\, d\, b^{E\setminus\delta}_{\varepsilon_{y_0}}(y)$ (Theorem XV,6, Corollary). Same for $R_1^{E\setminus\delta}(X)$. It is therefore sufficient to see that $\sup_\delta \int K(X,y) db^{E\setminus\delta}_{\varepsilon_{y_0}}(y) = 1$ (see Chap II, § 4) or even $R^{E\setminus\delta_n}_{K_X}(y_0) \longrightarrow \hat{R}^E_{K_X}(y_0)$ for a suitable δ_n ($\cap \delta_n = \{X\}$). Such a limit property is well known (Chap VI, § 10).

4. Reduced functions; minimal and non-minimal points.

Lemma XV, 10. If $e \subset \Omega$, v superharmonic ≥ 0 on Ω with representation $v(y) = \int K'(x,y)\, d\mu_v(x)$ ($\mu \geq 0$ on $\hat{\Omega}$ supported by $\Omega \cup \Delta_1$), then

$$\hat{R}^e_v(z) = \int \hat{R}^e_{K'(x,y)}(z)\, d\mu_v(x) .$$

Proof. We know (Chap. VI) that

$$\hat{R}^e_v(z) = \int v(y)\, d\, b^e_{\varepsilon_z}(y).$$

Hence it is

$$\int \left(\int K'(x,y)\, d\, b^e_{\varepsilon_z}(y) \right) d\mu_v(x) \text{ or } \int \hat{R}^e_{K'(x,.)}(z)\, d\mu_v(x).$$

Theorem XV, 11 (L. Naïm [1])(47). If u is harmonic ≥ 0, $E \subset \Omega$, then the condition $R^E_u = u$ is satisfied iff the borel set $\mathcal{E}_E$ of Δ_1 where E is thin has a μ_u-measure zero.

---

(47) We follow the proof of Gowrisankaran [1] which precises also $\mathcal{E}_E$.

Proof. Note that $\mathcal{E}_E$ is defined by $\hat{R}^{E'}_{K_X}(y_0) < 1$ ($E' = E \setminus \omega_0$ neighbourhood of y_0)

or
$$\int K_X \, d\, b^{E'}_{\varepsilon_{y_o}} < 1 \quad (X \in \Delta_1).$$

Suppose $\mu_u(\mathcal{E}_E) = 0$. As

$$\hat{R}^E_u(y) = \int \hat{R}^E_{K_X}(y)\, d\mu_u(X) \quad \text{and}$$

$$\hat{R}^E_{K_X} = K_X \quad (\forall X \notin \mathcal{E}_E), \text{ we get}$$

$$\hat{R}^E_u(y) = \int K_X(y)\, d\mu_u(X) = u(y).$$

Conversely, suppose $\hat{R}^E_u = u$; introduce a countable base of regular domains δ_i on Ω; any $X \in \Delta_1$ where E is thin satisfies $\hat{R}^E_{K_X}(y) < K_X(y)$ for one y in one δ_i; therefore, for all points of this δ_i; hence $\int \hat{R}^E_{K_X}(y)\, d\rho^{\delta_i}_y < K_X(y)$ $(\forall y \in \delta_i)$. In other words, $\mathcal{E}_E$ is contained in the countable union of the sets α_i of all X satisfying the previous inequality. Let us see that outer $\mu_u{}^*(\alpha_i) = 0$. As

$$\int \left(\int \hat{R}^E_{K_X}(x)\, d\rho^{\delta_i}_y(x) \right) d\mu_u(X)$$
$$= \int \left(\int \hat{R}^E_{K_X}(x) d\mu_u(X) \right) d\rho^{\delta_i}_y(X)$$
$$= \int \hat{R}^E_u(x)\, d\rho^{\delta_i}_y(x)$$
$$= \int u\, d\rho^{\delta_i}_y(x) = u(y) \quad (\forall y \in \delta_i).$$

We get $\int \left(K_X(y) - \int \hat{R}^E_{K_X}\, d\rho^{\delta_i}_y\right) d\mu_u(X) = 0$. Hence $\mu_u(\alpha_i) = 0$.

Remark. $\mathcal{E}_E$ is even the restriction of Δ_1 with a compact set of Δ. Because the set $\left\{X \in \Delta, \int_{\Omega_n} K_X \, d\, b^{E'}_{\varepsilon_{y_o}} \leq 1 - \frac{1}{p}\right\}$ is compact (domains Ω_n increasing, $\Omega_n \subset \bar{\Omega}_n \subset \bar{\Omega}$; $\bigcup \Omega_n = \Omega$).

Corollary. For any u harmonic $\geqslant 0$ and $E \subset \Omega$

(3)
$$\hat{R}_u^E = \int K_X \, d\mu_u' + \int \hat{R}_{K_X}^E \, d\mu_u''$$

where μ_u', μ_u'' are the restrictions of μ_u on $\Delta_1 - \mathscr{E}_E$ and $\mathscr{E}_E$.
The first term is the greatest harmonic minorant of $\hat{R}_u^E$, the second one is a potential.

In fact $X \in \Delta_1 - \mathscr{E}_E \Rightarrow \hat{R}_{K_X}^E = K_X$, hence the decomposition.

Now, if u' is harmonic > 0, $\leq \hat{R}_u^E$, let us see that $u' = R_{u'}^E$; if not, there exists a superharmonic $v \geqslant u'$ on E and $< u'$ at some x; consider w superharmonic $\geqslant u$ on E and close enough to R_u^E at x; then $v + w - u'$ would be $\geqslant 0$ ($w \geqslant R_u^E \geqslant u'$), $\geqslant u$ on E, but $< R_u^E$ at x.
Then $\mu_{u'}(\mathscr{E}_E) = 0$; as it is $\leq \mu_u$, $\mu_{u'} \leq \mu_u'$; hence $u' \leq \int K_X \, d\mu_u'$.

Definition XV, 12. u harmonic $\geqslant 0$, in Ω is said to be associated to zero at $X \in \Delta$, if there exists a neighbourhood δ of X in $\hat{\Omega}$ such that on $\delta \cap \Omega$, $u = R_u^{\complement \delta \cap \Omega}$. Then the same is true for any $\delta' \subset \delta$.

Proposition XV, 13. If u harmonic $\geqslant 0$ is associated to zero at any point of $\omega \cap \Delta_1$, where ω is an open set of $\hat{\Omega}$, then for the associated measure μ_u of u, $\mu_u(\omega \cap \Delta_1) = 0$. Conversely, that implies that u is associated to zero at every point of $\omega \cap \Delta$.

Proof. For any $X_0 \in \Delta_1 \cap \omega$, there exists on $\hat{\Omega}$, an open $\omega_1 \subset \omega$, neighbourhood of X_0 such that $\hat{R}_u^{\complement \omega_1 \cap \Omega} = u$ on $\omega_1 \cap \Omega$, therefore on Ω (because it is true q.e on $\complement \omega_1$); as $\complement \omega_1$ is thin at every point of $\omega_1 \cap \Delta_1$, $\mu_u(\omega_1 \cap \Delta_1) = 0$. Thanks to a countable covering of $\omega \cap \Delta_1$, we deduce $\mu_u(\omega \cap \Delta_1) = 0$.
Easy converse property. As immediate application,

Theorem XV, 14. u harmonic > 0 on Ω is minimal iff it is associated to zero at any $X \in \Delta_1$ except one (or any $X \in \Delta$ except one which is minimal).

Therefore, all harmonic functions $\geqslant 0$ associated to zero at every point of Δ except one X_0 are proportional to K_{X_0}, but equal to zero if X_0 is not minimal.

(Principle of positive singularities, inspired from Bouligand [1]. See Deny [1] and Brelot [12]).

Characterizations of non-minimal points of Δ

Proposition XV, 15. $X \in \Delta$ is not minimal if

a) There exists a neighbourhood δ of $X \in \Delta$ on $\hat{\Omega}$ such that $\hat{R}_{K_X}^{\delta \cap \Omega} \neq K_X$. Or also

b) there exists a measure $\mu \geqslant 0$ on Ω such that

$$(4) \qquad \int K(X,y)\, d\mu(y) < \liminf_{\substack{x \in \Omega \\ x \to X}} {}_{\mathcal{C}_m} \int K(x,y)\, d\mu(y).$$

Proof. The converse of (a) is close to the previous theorem.

As for (b), it implies X is not minimal because of formula (2).

If now X is not minimal, we use an open neighbourhood δ of (a) and the beginning of the argument of theorem XV, 6; that gives a measure satisfying (b).

Remark. Let us consider on $\hat{\Omega}$ (with topology $\mathcal{C}_m$), the cone Φ of functions of th. XV, 7. With the corresponding fine topology, Ω is strongly unthin at any $X \in \Delta_1$ (thanks to th. XV, 9). Moreover it is strongly thin at any $X \in \Delta - \Delta_1$. The thinness is contained in (b), the strong thinness is proved as in th. XV, 9.

Exercise. In this fine topology, $\Delta - \Delta_1$ is open and any subset of Δ_1 is closed.

Exercise. The minimal fine topology on $\Omega \cup \Delta_1$ is induced by a topology on $\hat{\Omega}$ such that $\forall X \in \Delta_1$, the sets of $\mathcal{G}_X$ plus X form a base of neighbourhood of X, and $\forall X \in \Delta - \Delta_1$, the neighbourhoods are those with $\mathcal{C}_m$.

5. Statistical thinness and Choquet property. (See Brelot [27]).

Theorem XV, 16. Given $e \subset \Omega$, $\alpha \subset \Delta_1$, h harmonic > 0 on Ω, whose associated measure on Δ_1 is μ_h, the h-statistical thinness on α (statistical property).

S.P. "e is thin on α except on a set of μ_h-measure zero" is equivalent to: "for neighbourhoods ω of α in $\hat{\Omega}$, the family $\{\omega \cap e\}$ is h-vanishing, i.e. $\inf \hat{R}_h^{\omega \cap e} = 0$. Same result with fine neighbourhoods of α in $\Omega \cup \Delta_1$.

Proof. Let us start from S.P. and decompose α in α_0 where e is unthin and α_1 where e is thin.

As $\mu_h(\alpha_0) = 0$, $\bar{\mathcal{D}}_{1,h}^{\alpha_0} = 0$; there exists v superharmonic > 0, such that $\liminf \frac{v}{h} \geqslant 1$ on α_0 and $v(y_0) < \varepsilon$; $\frac{v}{h} > 1 - \varepsilon$ on ω_0 open, intersection with Ω of ω open of $\hat{\Omega}$, containing α_0; hence $\frac{v}{1-\varepsilon} \geqslant h$ on ω_0, $R_h^{\omega \cap \Omega}(y_0) < \frac{\varepsilon}{1-\varepsilon}$. As for α_1, observe that the set where e is thin, is the intersection with Δ_1 of a K_σ-set on Δ (see th. XV, 11, Remark). Therefore, it is sufficient to prove the vanishing property for a compact set β of points of $\Delta - \Delta_1$ or where e is thin. Consider ω open on $\hat{\Omega}$, $\omega \supset \beta$ such that

$$\mu(\bar{\omega} \cap \Delta) < \mu(\beta) + \varepsilon \ (\bar{\omega}, \text{ closure in } \hat{\Omega})$$

and an increasing compact K_n such that $\bigcup K_n = \Omega$; $\hat{R}_h^{e \cap \omega}$ has a greatest harmonic minorant which is lim $\hat{R}_h^{e \cap \omega \cap \complement K_n}$ and also $\int_E K_X \, d\mu_h$, where E is the set of Δ_1 where $e \cap \omega$ is unthin (Corollary theorem XV, 11). Hence $\hat{R}_h^{e \cap \omega \cap \complement K_n}(y_0) \longrightarrow \mu(E)$. But $E \subset \bar{\omega} - \beta$, hence $\mu(E) < \varepsilon$ and hence the wanted vanishing property.

Conversely, consider a fine neighbourhood ω of α such that $\hat{R}_h^{e \cap \omega}(y_0) < \varepsilon$. The greatest harmonic minorant is $\int_E K_X \, d\mu_h$ with the same meaning of E; but $\alpha_0 \subset E$, hence

$$\overline{\int_{\alpha_0}} d\mu_h \leq \int_E d\mu_h \leq \hat{R}_h^{e \cap \omega}(y_0) < \varepsilon ;$$

then $\mu_h(\alpha_0) = 0$.

<u>Complements</u> (Indications)

1). Equivalence with the existence of U superharmonic > 0 such that $U/h \longrightarrow \infty$ on $\alpha \cap \bar{e}$ (or the same with the fine topology).

2) h may be replaced by V superharmonic > 0, with the use of μ_V associated to V and equal on Δ to μ_h, where h is the greatest harmonic minorant of V.

3) Comparing the similar study in Ω (Chap VII), we conclude for $\alpha \subset \Omega \cup \Delta_1$, but $e \subset \Omega$ and V superharmonic > 0: the similar statistical property with exceptional V-polar set (i.e. V statistical thinness) is equivalent to the V-vanishing property of $\{\omega \cap e\}$ (ω, a neighbourhood of α in $\hat{\Omega}$ or in

fine topology) and to the existence of $U \geqslant 0$ superharmonic such that $U/V|_e$ tends to $+\infty$ (where it has a meaning) at the points of $\alpha \cap \bar{e}$ (resp. $\alpha \cap \tilde{e}$ with fine convergence).

Choquet property. In the general theory (Chap. IV), we may call Choquet property relative to a set e and a weight $\wp(e)$, the existence of a closed set $e'_0 \subset \tilde{e}_0$ such that $\wp(\tilde{e}_0 - e'_0) < \varepsilon$ arbitrarily chosen.

Exercise. In the previous classical theory, let us denote $\overset{*}{R}{}_\varphi^e$, the lower envelope of the functions, hyperharmonic $\geqslant 0$ majorizing $\varphi \geqslant 0$ on $e \cap \Omega$ and also on $\Omega \cap \omega$ for one neighbourhood ω of e in $\hat{\Omega}$, then for a superharmonic $V > 0$, the weight $\int \hat{R}_\varphi^e \, d\rho_{y_0}^\delta$ ($d\rho_{y_0}^\delta$, harmonic measure of δ at $y_0 \in \delta$) gives the Choquet property to every $e \subset \Omega \cup \Delta_1$ such that $e \cap \Delta_1$ is μ_V-measurable.

6. Case of a ball or a half-space of $\mathbb{R}^n$

The Martin space is the Euclidean closure of the domain. It is possible and interesting to precise the previous study as L. Naïm [1] did, and to give for instance a criterion of thinness similar to the classical one of Wiener.

A more detailed direct and different study had been developed previously by Mme Lelong [1] in the half-space, but the definitions (one of thinness by the Wiener type criterion) and methods were not adaptable to a general Martin space and are therefore left aside here (see however Chap. XVII) It would be important to deepen the results and their comparison with the general theory.

Complements of this chapter (for instance about subspaces) may be found in Martin [1], Brelot [12], Naïm [1] and in recent axiomatic researches.

CHAPTER XVI

Classical Martin Boundary

Dirichlet Problem and Boundary Behaviour

1. A general Dirichlet problem (Brelot [16]).

We consider a Green space Ω, dense in a compact metrizable $\overline{\overline{\Omega}}$ (with a boundary $\Delta = \overline{\overline{\Omega}} - \Omega$) and a fixed harmonic function $h > 0$ on Ω.

Lemma XVI, 1. If v is superharmonic on Ω and satisfies $\liminf_{\substack{y \to X \in \Delta \\ y \in \Omega}} \frac{v(y)}{h(y)} \geq \lambda$ ($\forall X \in \Delta$ topology of $\overline{\overline{\Omega}}$), then $\frac{v}{h} \geq \lambda$; (it is sufficient to suppose $\lambda > -\infty$).

Proof. If not, $\frac{v}{h}$ with a lower semi-continuous continuation on $\overline{\overline{\Omega}}$, attains its minimum $k < \lambda$ on Ω, $v - kh$ superharmonic ≥ 0 on Ω, but zero at a point would be zero. Then $\frac{v}{h} = k < \lambda$. This is a contradiction.

General Envelopes. Given a real f on Δ, we consider the family Φ of the hyperharmonic functions v on Ω satisfying $\liminf \frac{v}{h} \geq f(X)$ ($\forall X \in \Delta$). Note that this condition "$> -\infty$" is equivalent to "$\frac{v}{h}$ lower bounded". The lower envelope $\overline{\mathcal{D}}_{f,h}$ is equal to $+\infty$, $-\infty$ or harmonic (and obviously increasing in f). In case $f = \varphi_e$, indicator of $e \subset \Delta$, we write also h_e. When $h_e = 0$, we say e is h-negligible; "h-almost everywhere" (h-a-e) will mean "except on a h-negligible set of Δ". Note $\overline{\mathcal{D}}_{f,h}$ is preserved by changing f on a h-negligible set. Therefore, f may not be defined on such a set.

We define $\underline{\mathcal{D}}_{f,h}$ as $-\overline{\mathcal{D}}_{-f,h}$ and the lemma implies $\underline{\mathcal{D}}_{f,h} \leq \overline{\mathcal{D}}_{f,h}$. In case both envelopes are equal and finite (then harmonic), we say f is h-resolutive and the common envelope is denoted $\mathcal{D}_{f,h}$ (harmonic solution of the h-Dirichlet problem)(48). It is obvious that 1 is h-resolutive and $\mathcal{D}_{1,h} = h$. This h-resolutivity is preserved by linear combination and uniform convergence.

We will develop a theory similar to the classical one for a bounded domain of $\mathbb{R}^n$ (given in Brelot [24]; see a general frame in Brelot [18 - 19]).

---

(48) With the language of the axiomatic theory (Brelot [19], $\mathcal{D}_{f,h}/h$ would be called solution of a Dirichlet problem relative to the "h-harmonic functions" (quotients by h of harmonic functions) and the given boundary function f.

Lemma XVI, 2. If f_n is increasing and $\mathfrak{D}_{f_n,h} > -\infty$, then $\overline{\mathfrak{D}}_{f_n,h}$ is increasing and tends to $\overline{\mathfrak{D}}_{\lim f_n,h}$.

Same argument as in the classical case or in a general frame (Brelot [20], [25]). As a particular case, if $e_n \uparrow$, $\bigcup e_n = e, h_{e_n} \uparrow, h_{e_n} \to h_e$.

Theorem XVI, 3. Suppose that any finite real-valued continuous function on Δ is h-resolutive (condition $\mathcal{Q}_h$). Then the linear form of φ, $\mathfrak{D}_{\varphi,h}(y)$ may be written $\int \varphi \, d\nu_h^y$ (ν_h^y unique positive Radon measure on Δ) and for any f: $\overline{\mathfrak{D}}_{f,h}(y) = \overline{\int} f \, d\nu_h^y$. Hence $h_e(y)$ is the outer ν_h^y-measure of e. The h-resolutivity of f is equivalent to the ν_h^y-integrability (independant of y) and in that case $\mathfrak{D}_{f,h}(y) = \int f \, d\nu_h^y$.

Same proof as in the classical or general frame.

We will need also:

Lemma XVI, 4. $h_e = \operatorname{Inf}_{\alpha} R_h^{\alpha \cap \Omega}$ (denoted h_e'), α an open neighbourhood of e on $\overline{\overline{\Omega}}$.

Proof. It is obvious that $h_e \leq R_h^{\alpha \cap \Omega}$; now if $v \geqslant 0$ superharmonic satisfies $\liminf \frac{v}{h} \geqslant \varphi_e$ at any boundary point, $\frac{v}{h}$ continued by lower semi-continuity is $> 1-\varepsilon$ on an open set β of $\overline{\overline{\Omega}}$, containing e; $\frac{v}{1-\varepsilon} > h$ on $\beta \cap \Omega$; $\frac{v}{1-\varepsilon} \geqslant R_h^{\beta \cap \Omega}$ and hence $h_e \geqslant (1-\varepsilon) R_h^{\beta \cap \Omega} \geqslant (1-\varepsilon) h_e'$. As ε was arbitrarily chosen, $h_e \geqslant h_e'$.

Corollary 1. There exists a sequence α_n of open neighbourhoods of e on $\overline{\overline{\Omega}}$, decreasing such that $h_e = \operatorname{Inf}_n R_h^{\alpha_n \cap \Omega}$.

2. If γ is any neighbourhood of e on Δ, $h_e = \inf_{\gamma} h_{\gamma}$.

Lemma XVI, 5. $(h_e)_e = h_e$ ($\forall e \subset \Delta$).

Having chosen an open neighbourhood α of e on $\overline{\overline{\Omega}}$, we introduce α_n decreasing $\subset \alpha$ of the previous type (Cor. 1). Then

$\hat{R}^{\alpha \cap \Omega}_{\hat{R}_h^{\alpha_n \cap \Omega}} = \hat{R}_h^{\alpha_n \cap \Omega}$ (49). The first member tends to $\hat{R}_{h_e}^{\alpha \cap \Omega}$ (50). Now $\inf_{\alpha} \hat{R}_{h_e}^{\alpha \cap \Omega}$ is $(h_e)_e$. Hence the wanted equality.

Many complements and further developments may be found in Brelot [16] and L. Naïm [1]. Study of $\mathcal{a}_h$, variation of h, of Ω in a fixed Green space, of the kind of compactification or of boundary, of the behaviour of functions at the boundary etc. We shall mention some points: Without $\mathcal{a}_h$, the application $e \rightsquigarrow h_e(y)$ for e compact is a Choquet's capacity (even alternate of infinite order) and it becomes additive (i.e. defines a measure whose outer measure will be h_e for any set e) iff $\mathcal{a}_h$ is satisfied. In case of $\mathcal{a}_h$, a basis of filter $\mathfrak{F}$ converging to $X \in \Delta$ is said to be h-regular if $\mathcal{D}_{f,h}/h \underset{\mathfrak{F}}{\longrightarrow} f(X)$ ($\forall f$ finite continuous). This is equivalent to

a) $\limsup\limits_{\mathfrak{F}} \dfrac{\overline{\mathcal{D}}_{f,h}}{h} \leq \limsup f$ at X ($\forall f$ upper bounded) and to

b) $\dfrac{h_e}{h} \underset{\mathfrak{F}}{\longrightarrow} 0$ ($\forall e$ compact $\not\ni X$);

(a) and (b) are equivalent without $\mathcal{a}_h$ and may define the h-regularity of $\mathfrak{F}$ in the general case.

A boundary point $X \in \Delta$ is said to be h-regular if the neighbourhoods intersect Ω according to a h-regular $\mathfrak{F}$. With this notion, some classical results with the Euclidean boundary may be extended; for instance, the lim inf, denoted $\mathcal{L}_h^u(X)$ of $\frac{u}{h}$ (u superharmonic, $\frac{u}{h}$ lower bounded) at $X \in \Delta$ (X, h regular) is equal to $\liminf\limits_{\substack{Y \to X \\ Y \in \Delta - E}} \mathcal{L}_h^u(Y)$ where E is h-negligivle. But we do not know if the set of the h-irregular boundary points is h-negligible.

---

(49) because of the following remark: for open sets $\beta \subset \beta' \subset \Omega$, $\hat{R}^{\beta'}_{\hat{R}_v^{\beta}} = \hat{R}_v^{\beta}$.

The inequality $\leq$ is obvious. The first member majorizes v on β and therefore the second member (same result for any sets $\beta \subset \beta''$).

(50) because of the general formula (8) Chap VI p. 51.

Let us mention too that, in the general frame of the Dirichlet problem, metrizable but complete and perhaps non-compact spaces and boundaries have been studied with boundary conditions expressed by means of filters. Interesting examples are given by completion from a metric, compactible with the topology of Ω and by bases of filters which may be defined in a particular case, by the ends of lines, for example the Green lines (Brelot and Choquet [1], Brelot [15], Ohtsuka, M. Arsove and G. Johnson).

2. Fundamental case of the Martin space $\hat{\Omega}$.

We will go further with $\bar{\bar{\Omega}} = \hat{\Omega}$. (see Brelot [17]).

Lemma XVI, 6. Suppose X is minimal (i.e. $X \in \Delta_1$). If $X \in e$, $(K_X)_e = K_X$; if not, $(K_X)_e = 0$.

Proof. For any open neighbourhood α of e on $\hat{\Omega}$, $\alpha \cap \Omega$ is unthin at $X \in e$ (lemma XV, 1). Therefore, $R_{K_X}^{\alpha \cap \Omega} = K_X$ (definition of thinness). Hence $(K_X)_e = K_X$.

If $X \notin e$, let us prove more generally that $(K_X)_{\Delta - X} = 0$. We may use the existence of a neighbourhood E of $\Delta \setminus X$ on $\hat{\Omega}$ such that $E \cap \Omega$ is thin at X. (see lemma XV, 2). Then $R_{K_X}^{E\cap\Omega} < K_X$, and hence $(K_X)_{\Delta - X} < K_X$. Now, it is sufficient to remark that for any $e \subset \Delta$, K_X minimal satisfies $(K_X)_e = 0$ or K_X. Because $(K_X)_e \leq K_X$ minimal, $(K_X)_e = l\, K_X$ (l real ≥ 0) $((K_X)_e)_e = l(K_X)_e$, hence $(K_X)_e = l(K_X)_e$; therefore, $(K_X)_e$ is zero or $l = 1$.

Lemma XVI, 7. $h_e(y) = \int_e K_X(y)\, d\mu_h(X)$, i.e. $\int \varphi_e \cdot K_X\, d\mu_h$ (φ_e, the characteristic function of e) where μ_h is the measure of the Martin representation of h (see Theorem XIV, 4). In case e is Borel (or even μ_h-measurable), denote by μ_h^e, the restriction of μ_h on e. Then $h_e(y) = \int K_X\, d\mu_h^e(X)$, $h_e(y_0) = \int_e d\mu_h$.

Proof. Suppose, first e is compact. Consider the open neighbourhoods α of e on $\hat{\Omega}$.

$$\hat{R}_h^{\alpha\cap\Omega}(y) = \int \hat{R}_{K_X}^{\alpha\cap\Omega}(y)\, d\mu_h(X) \text{ (lemma XV, 10).}$$

By taking a sequence α_n of such α, decreasing with intersection e, we see that

$\hat{R}_h^{\alpha_n \cap \Omega}$ and $\hat{R}_{K_X}^{\alpha_n \cap \Omega}$ tend respectively to h_e and $(K_X)_e$. Hence

$h_e(y) = \int (K_X)_e \, d\mu_h(X) = \int_e (K_X) \, d\mu_h(X)$.

For any open e, we consider an increasing sequence of compact sets $e_n(\bigcup e_n = e)$ and we get the same formula. For any e, we consider any open neighbourhood j on Δ; $h_e = \inf_j h_j$; for a fixed y, $\int_j K_X(y) \, d\mu_h$ is the $K_X(y)d\mu_h$ measure of j and the $\inf_j$ is the corresponding outer measure of e; i.e.

$= \overline{\int} \varphi_e K_X(y) \, d\mu_h(X)$.

Note that $\Delta - \Delta_1$ is always h-negligible.

Corollary. If e_1, e_2 are disjoint Borel sets on Δ, $(h_{e_1})_{e_2} = 0$.

Because $(h_{e_1})_{e_2}(y_0) = \int_{e_2} d\mu_h^{e_1} = 0$.

Lemma XVI, 8. If e is a Borel set (or even μ_h-measurable), the characteristic function φ_e of e is h-resolutive. As $(h_e)_{\Delta - e} = 0$, for a suitable open neighbourhood α of $\Delta - e$ on $\hat{\Omega}$, we have

$\hat{R}_{h_e}^{\alpha \cap \Omega}(y_0) < \varepsilon$ (arbitrarily chosen); $h_e - \hat{R}_{h_e}^{\alpha \cap \Omega}$ is $\geqslant 0$ and zero on $\alpha \cap \Omega$; this subharmonic function on Ω is $\leqslant h_e = \overline{\mathcal{D}}_{\varphi_e, h}$ and the quotient by h tends to zero at every point of $\Delta - e$; it minorizes $\underline{\mathcal{D}}_{\varphi_e, h}$. Hence $\overline{\mathcal{D}}_{\varphi_e, h} - \underline{\mathcal{D}}_{\varphi_e, h}$ is arbitrarily small at y_0, is zero at y_0 and hence on Ω.

Theorem XVI, 9 (Martin space). Any finite continuous function φ on Δ is h-resolutive and the $d\nu_h^y$ of the general theory is now $K_X(y) \, d\mu_h(X)$.

Proof. In fact, we may approach φ upto ε by a finite sum of functions of the type $\lambda_i \varphi_{e_i}$ (e_i Borel set, λ_i real) which are h-resolutive. The comparison of $h_e(y)$ given in theorem XVI, 3 and in lemma XVI,6 gives the wanted relation between ν_h^y and μ_h. Note that h-negligible now means of "μ_h - measure zero".

Remark 1. A similar but longer development was given first by Brelot [17] without the use of the previous Martin integral representations, but using a rough elementary integral representation. That led to the expression $\mathcal{D}_{f,h} = \int f.K_X\, d\nu_h^{y_o}$ where $\nu_h^{y_o}$ is supported by Δ_1; this contains $\mathcal{D}_{1,h} = h = \int K_X d\nu_h^{y_o}$ which is the Martin representation.

Remark 2. The lemma XVI, 4 leads to the study of $R_h^{e_i \cap \Omega}$ (in $\hat{\Omega}$) for a decreasing directed set $\{e_i\} \subset \Omega$; but even when e_i is open and $\bigcap e_i \cap \Omega = \emptyset$, the inf is not $h_{\bigcap e_i}$; when $\bigcap e_i \cap \Omega = \emptyset$, it is inf h_{A_i} where A_i is the set of the points of Δ_1 where $e_i \cap \Omega$ is not thin. A deeper and more general study is developed in Brelot [30].

Exercise. The functions $\mathcal{D}_{f,h}(f \geqslant 0)$ are characterized as limits of increasing sequences of non-negative harmonic functions whose quotients by h are (independently) bounded (Parreau [1], Brelot [17]).

3. Other characterisations of the envelopes (Martin case).

Lemma XVI, 10 (Naim [1]). Minimum principle. Given h harmonic > 0, u superharmonic and $\frac{u}{h} \geqslant K$ (constant $\leqslant 0$), suppose there exists $\forall X \in \Delta_1 - e$, e such that inner μ_h measure of $e = 0$, an unthin set E_X s.t. $\liminf\limits_{\substack{y \to X \\ y \in E_X}} \frac{u(y)}{h(y)} \geqslant 0$. Then $u \geqslant 0$.

Proof. Consider $u_1 = \inf(u,0)$ and suppose $u_1 \neq 0$; there exists a greatest harmonic minorant $u_1' \geqslant Kh$. The set $E_\varepsilon = \{x \mid u_1(x) \geqslant -\varepsilon h(x)\}$ is also the set $\{x \mid u \geqslant -\varepsilon h\}$ and contains for every $X \in \Delta_1$, $X \notin e$ the intersection of E_X with a neighbourhood of X; the Borel set of Δ_1 where E_ε is thin, is therefore a part of e, and hence has μ_h-measure zero and also $\mu_{-u_1'}$ measure zero. Hence $R^E_{-u_1'} = -u_1'$ (theorem XV, 11). As $u_1 = u_1' + v$ (v potential $\geqslant 0$), $-u_1' \leqslant \varepsilon h + v$ on E_ε. Therefore $R^{E_\varepsilon}_{-u_1'} \leqslant \varepsilon h + v$ on Ω where ε is arbitrary; hence $u_1' = 0$. This is a contradiction.

Theorem XVI, 11 (Naïm [1], Doob [3]). For any real function f on Δ, let us consider the hyperharmonic functions u such that $\frac{u}{h}$ (h fixed harmonic) is lower bounded and satisfies one of the following conditions:

a) There exists $\forall$ u, at any $X \in \Delta_1$, a set E_X^u unthin at X such that $\liminf\limits_{y \in E_X^u,\, y \to X} \frac{u}{h} \geqslant f(X)$.

b) $\liminf\limits_{\text{fine at } X} \frac{u}{h} \geqslant f(X) \quad \forall \quad X \in \Delta_1$.

c) $\limsup_{\text{fine at } X} \frac{u}{h} \geq f(X) \quad \forall\, X \in \Delta_1$. (Doob)

In each case, the lower envelope of these u is $\overline{\mathcal{B}}_{f,h}$. Same result if we suppose that these conditions are satisfied μ_h a.e.

Proof. The new envelopes are obviously $\leq \overline{\mathcal{B}}_{f,h}$. We need the opposite inequality. As $\frac{u}{h}$ is lower bounded by a constant K, it is sufficient to consider the case with $f' = \sup(f,K)$, because $\overline{\mathcal{B}}_{f',h} \geq \overline{\mathcal{B}}_{f,h}$. We will therefore suppose f lower bounded, even for instance ≥ 0. Suppose $\overline{\mathcal{B}}_{f,h}$ is finite. There exists a function $\varphi \geq f$, ν_h^y or μ_h-integrable such that $\mathcal{B}_{\varphi,h} = \overline{\mathcal{B}}_{f,h}$ and $\underline{\mathcal{B}}_{\varphi - f,h} = 0$ ($\varphi - f = 0$ when f is infinite) - property of the upper integral. This implies that the set where $\varphi - f > \varepsilon > 0$ has inner μ_h-measure zero. Now, if v hypoharmonic satisfies: $\frac{v}{h}$ upper bounded, $\limsup \frac{v}{h} \leq \varphi$ at any $X \in \Delta_1$, we see that $u - v$, hyperharmonic with $\frac{u-v}{h}$ lower bounded, satisfies in case

a) $\liminf_{y \in E_X^u,\, y \to X} \frac{u - v + \varepsilon h}{h} \geq 0 \quad \forall\, X \in \Delta_1$, except on a set of inner μ_h-measure zero.

Hence $u - v \geq -\varepsilon h$, $u \geq v$, $u \geq \underline{\mathcal{B}}_{\varphi,h} = \mathcal{B}_{\varphi,h} = \overline{\mathcal{B}}_{f,h}$.

In the case of (b) and (c) $\frac{u(y)}{h(y)}$ has the limit $f(X)$ on a suitable $E_X'^u$, unthin at $X \in \Delta_1$, because of the interpretation of the fine topology on $\Omega \cup \Delta_1$ in theorems XV, 9 and III, 3. Therefore $u - v$ satisfies the same previous condition with $E_X'^u$ instead of E_X^u and we conclude $u \geq \overline{\mathcal{B}}_{f,h}$.

If $\overline{\mathcal{B}}_{f,h} = +\infty$, let us consider $f_n = \inf(f, n)$; $\overline{\mathcal{B}}_{f_n,h}$ is finite and tends to $\overline{\mathcal{B}}_{f,h}$ as $n \to \infty$. As $u \geq \overline{\mathcal{B}}_{f_n,h}$, $u \geq \overline{\mathcal{B}}_{f,h} = +\infty$.

Remark. As a consequence in case (c), if v is superharmonic > 0, the set of Δ_1 where fine $\limsup \frac{v}{h} = +\infty$ is h-negligible (direct proof in Naim [2]).

4. Boundary behaviour of harmonic and superharmonic functions: Global results thanks to the (minimal) fine topology (Martin case).

Because of the success of the minimal fine topology, the classical Fatou theorem on non-tangential boundary limit for bounded harmonic functions in a circle and further improvements of Calderon, Stein etc. have inspired a similar study by Doob, by changing these limits with fine limits, but now in a general Martin space.

The background was the previous study.

<u>Theorem XVI, 12</u> (Doob [3] and [4]). <u>Given</u> h <u>harmonic</u> > 0 <u>and</u> f <u>h-resolutive on</u> Δ, $\frac{\mathcal{H}_{f,h}}{h}$ <u>has fine limit</u> f μ_h <u>a.e. on</u> Δ_1.

<u>Proof</u>. Denote by f', the fine lim sup of $\frac{\mathcal{H}_{f,h}}{h}$ on Δ_1 and $f_1 = \sup(f, f') \geqslant f$. If u satisfies the conditions of the previous theorem (case c), $u \geqslant \mathcal{H}_{f,h}$. Then, by the same argument $u \geqslant \overline{\mathcal{H}}_{f_1,h}$; hence for the lower envelope of these u, $\mathcal{H}_{f,h} \geqslant \overline{\mathcal{H}}_{f_1,h}$ then $\mathcal{H}_{f,h} = \overline{\mathcal{H}}_{f_1,h} = \underline{\mathcal{H}}_{f_1,h}$; this implies $f = f_1$ μ_h a.e. i.e. $f' \leq f$ μ_h a.e. Same result with $-f$; so we get fine lim inf $\frac{\mathcal{H}_{f,h}}{h} \geqslant f$ μ_h a.e.

<u>Remark</u>. See another proof in Naim [2], using instead of theorem XVI, 10 (case c), the previous remark.

<u>Theorem XVI, 13</u> (Naim [1]). <u>If</u> v <u>is a potential</u>, h <u>harmonic</u> > 0, $\frac{v}{h}$ <u>has on</u> Δ_1, <u>a fine limit zero</u> μ_h <u>a.e.</u>

<u>Proof</u>. We have to see that the set of $X \in \Delta_1$ where fine lim sup $\frac{v}{h} > 0$ is h-negligible. Let us prove the same for: fine lim sup $\frac{v}{h} > \varepsilon > 0$. This set is contained in the set of $X \in \Delta_1$ where $E = \left\{x \mid \frac{v(x)}{h(x)} > \varepsilon\right\}$ is unthin; now $\hat{R}_h^E \leq \frac{v}{\varepsilon}$, therefore is a potential and this implies, according to corollary of theorem XV, 11, that the set of Δ_1 where E is unthin is h-negligible.

<u>Corollary 1</u>. Given h, δ open on $\hat{\Omega}$, u harmonic > 0 such that $\mu_u(\delta \cap \Delta) = 0$, then $\frac{u}{h}$ has fine limit 0, μ_h a.e. on $\delta \cap \Delta_1$.

We introduce on $\hat{\Omega}$, δ_0 open $\subset \bar{\delta}_0 \subset \delta$. As the set of Δ_1 where $\delta_0 \cap \Omega$ is unthin is contained in $\delta \cap \Omega$, therefore is u-negligible, $R_u^{\delta_0 \cap \Omega}$ is a potential (theorem XV, 11; corollary) but is equal to u on $\delta_0 \cap \Omega$. Hence the wanted result for δ_0, therefore for δ.

<u>Application</u>. If u harmonic > 0 is associated to 0 at $X \in \Delta_1$, $\frac{u}{h}$ has a fine limit 0 μ_h a.e. in a neighbourhood of X. Particular case of K_{X_0} ($X_0 \in \Delta_1$, $X \in \Delta_1$, $X \neq X_0$).

Because $\mu_u(\delta \cap \Delta_1) = 0$ for a suitable open neighbourhood δ of X (Prop. XV, 13).

Corollary 2. If in $\hat{\Omega}$, δ is open, $\hat{R}_h^{\complement\delta\cap\Omega}/h$ has a fine limit 0 μ_h a.e. on $\delta\cap\Delta_1$.

In fact, according to theorem XV, 11, corollary, $\hat{R}_h^{\complement\delta\cap\Omega}$ is the sum of a potential and of a harmonic function u such that $\mu_u(\delta\cap\Delta_1) = 0$, because $\complement\delta\cap\Omega$ is thin on $\delta\cap\Delta_1$.

Lemma XVI, 14. Any harmonic function $u \geqslant 0$ is the sum of a $\theta_{f,h}$ (for a suitable μ_h-integrable f) and of a harmonic v such that inf (v,h) is a potential [Parreau (case h = 1) [1], then Naim [1]).

Proof. In fact μ_u is the sum of a measure $f.\mu_h$ (f, μ_h-integrable) and of a measure ν singular relative to μ_h i.e. such that inf $(\nu, \mu_h) = 0$ (F. Riesz). In other words, u is the sum of $\int K_X \cdot f\, d\mu_h$ and of a harmonic $v \geqslant 0$ such that h and v have no common harmonic minorant $\geqslant 0$ except 0, i.e. inf (v,h) is a potential. (See other forms of the proof without measures in Parreau [1] and Naim [1]).

Lemma XVI, 15 (Naïm [1]). Given u and h harmonic > 0, if inf (u,h) is a potential, $\frac{u}{h}$ has a fine limt 0, μ_h a.e. on Δ_1.

Consider $E_\varepsilon = \left\{ x \,\Big|\, \frac{u(x)}{h(x)} > \varepsilon \right\}$ $(0 < \varepsilon < 1)$. $R_h^{E_\varepsilon}$ minorizes h and $\frac{u}{\varepsilon}$, therefore also $\frac{1}{\varepsilon}$ inf (u, h). It is a potential, therefore the set of Δ_1 where E_ε is unthin has μ_h-measure zero (theorem XV, 11, corollary); the same is true for the set where fine lim sup $\frac{u}{h} > 0$.

Remark. The converse is true (Naim [1]).

Fundamental theorem XVI, 16 (Doob [3], [4]). Given u superharmonic > 0, h harmonic > 0, $\frac{u}{h}$ has a fine (finite) limit μ_h a.e. on Δ_1.

Proof. In fact u = v (harmonic $\geqslant 0$) + a potential. The behaviour of the potential is given by theorem XVI, 13, the behaviour of v by the decomposition of lemma XVI, 14, by the lemma XVI, 15 and by the lemma XVI, 12 where f is μ_h-integrable, therefore finite μ_h a.e.

Extensions. Let us just mention:

a) Doob ([3], [4]) even proved that for u and h both superharmonic > 0, $\frac{u}{h}$ has a finite fine limit μ_h a.e. on $\Omega \cup \Delta_1$ (μ_h on $\hat{\Omega}$ associated to h by the Riesz-Martin representation).

b) For $h > 0$, but u superharmonic satisfying: $\frac{u}{h}$ lower bounded on a fine neighbourhood of each point of $A \subset \Delta_1$, then $\frac{u}{h}$ has a finite fine limit μ_h a.e. on A (Doob [4]).

Complements. α) Doob [6] studied also the boundary behaviour of the BLD functions already considered in Chap. IX; he proved there is a fine limit f μ_1 a.e. on Δ_1 and when the function u is harmonic, $u = \mathcal{D}_{f,1}$; and the Dirichlet integral may be evaluated in terms of f; moreover $f = 0$ (μ_1 a.e) iff u is of "potential type" i.e. limit of C^{∞}- BLD functions with compact supports (limit of a sequence q.e. and also according to the Dirichlet norm). See further results in Doob. Extension to h- BLD functions by Lumer-Naïm.

β) v harmonic is called "generalized conjugate" of u harmonic if $|\text{grad } v| \leq K\,|\text{grad } u|$ on Ω (K a constant). Now, v has a minimal fine limit on Δ_1 a.e. when u has (Doob [9], with slight extension).

5. Fine Dirichlet Problem and various types of regularity.

We do not know if the set of the non-h regular points (defined in the more general case § 1) is h-negligible even in the Martin case. But with another similar Dirichlet problem, that will become true for a corresponding regularity. We consider the functions u of theorem XVI, 10, case b and their lower envelope, which is equal to $\overline{\mathcal{D}}_{f,h}$; we consider also $\underline{\mathcal{D}}_{f,h} = -\overline{\mathcal{D}}_{-f,h}$; for the corresponding so called "fine h-Dirichlet problem", the h-resolutivity and the solution are the same as previously. But the corresponding fine h-regular boundary points will be defined by the weaker condition $\frac{\mathcal{D}_{f,h}}{h} \longrightarrow f(X)$ $(X \in \Delta)$ in the fine topology, $\forall f$ finite continuous on Δ. That leads immediately to the wanted satisfactory property of the non-fine h-regular points.

In fact, there exists a countable dense subset $\{f_i\}$ in the space of the finite continuous real f on Δ (with topology of the uniform convergence). If e_i is the exceptional set corresponding to f_i by the theorem XVI, 12, $\bigcup e_i$ is h-negligible and at any $X \in \Delta_1 - \bigcup e_i$, $\frac{\mathcal{D}_{f,h}}{h} \longrightarrow f(X)$ in the fine topology.

We will prove more and deepen the question by introducing new concepts (we will consider here only for $\hat{\Omega}$). See Brelot [17] and chiefly Naïm [1].

A filter $\mathfrak{F}$ on Ω converging in $\hat{\Omega}$ to $X \in \Delta_1$ is said to be <u>weak</u> h-regular, if there exists on Ω a superharmonic $v > 0$ such that $\frac{v}{h} \xrightarrow[\mathfrak{F}]{} 0$, and <u>strong</u> h-regular if moreover, $\inf \frac{v}{h} > 0$ outside every neighbourhood of X in $\hat{\Omega}$.

It is not obvious that each h-regularity is equivalent to the existence of such a v only in a neighbourhood of X; but it is easy to see that the h-regularity implies the weak one and is implied by the strong one; and also that the weak one is equivalent to $\frac{G_{y_0}}{h} \xrightarrow[\mathfrak{F}]{} 0$ (obviously independant of y_0) and the strong one to $R_h^{\complement(\delta \cap \Omega)}/h \longrightarrow 0$, for every neighbourhood δ of X.

If $\mathfrak{F}$ is defined by the intersection of Ω with the (<u>fine</u>) neighbourhoods of X, X is said respectively (<u>fine</u>) <u>weak h-regular</u> or (<u>fine</u>) strong h-regular.

Recall (th XV, 8) that $\frac{G_{y_0}}{h}$ has at any $X \in \Delta_1$, a fine lim sup equal to the lim sup in $\hat{\Omega}$, therefore the weak h-regular points are identical to the fine weak h-regular ones.

<u>Theorem XVI, 17</u> (Chiefly Naïm [1]). <u>On</u> Δ_1 <u>each of the following sets is h-negligible: sets of points which are</u>

a) <u>not fine weak</u> (<u>or weak</u>) h-<u>regular</u>,

b) <u>not fine</u> h-<u>regular</u>,

c) <u>not fine strong</u> h-<u>regular</u>.

<u>Proof</u>. The first part is given by Theorem XVI, 13, but is a consequence of the other ones. The second one was already proved and is also a consequence of the last one, we will prove now. We consider a countable base of open sets of $\hat{\Omega}$ and those δ_n intersecting Δ_1. $R_h^{\complement(\delta_n \cap \Omega)}/h$ tends finely to zero on $\delta_n \cap \Delta_1$ except on a h-negligible set e_n (according to theorem XVI, 13, corollary 2). Hence $\cup e_n$ is h-negligible and contains the non-fine strong h-regular points.

<u>Exercise</u>. In an open $\omega \subset \Omega \subset \hat{\Omega}$, we may define a h-Dirichlet problem by using the boundary of ω in $\hat{\Omega}$. (See Brelot [17]) even in the general frame of § 1. Same definitions of h-regularity. Now, the strong h-regularity of $\mathfrak{F}$ converging

to $X \in \Delta_1$ is equivalent to the h-regularity relative to $\omega \cap \Omega$, for any open neighbourhood of X.

<u>Remark on the use of uniform integrability</u>. With the Martin or minimal fine boundary, we may get better results than with the euclidean boundary (Chap IX, §8). Given a harmonic function u in Ω, it is a solution $\mathcal{D}_{f,1}$ iff u satisfies the uniform integrability property relative to the harmonic measures $\mu_{x_0}^{\omega_i}$ (ω_i relatively compact domain in Ω, $\omega_i \ni x_0$), then f is μ_1 - a.e, the function of fine limits.

And a superharmonic function with fine limit 0 μ_1 a.e is a potential iff it satisfies the previous uniform integrability property.

Easy extension to quotients $\frac{u}{h}$ (h fixed harmonic > 0).

6. <u>Applications and further study of the Dirichlet problem</u>. <u>Short indications</u>.

The previous concepts lead to a deeper study of the behaviour of the superharmonic functions in the neighbourhoods of boundary points. See Naim [1]. For example, at $X_0 \in \Delta$ not weak-h-regular,

a) If v is superharmonic > 0, $\frac{v}{h}$ has a fine limit $\Lambda_v \geq 0$ at X_0.

b) If $\mathfrak{F}$ converges to X_0, and $\frac{G_{y_0}}{h} \xrightarrow[\mathfrak{F}]{}$ lim sup $\frac{G_{y_0}}{h}$ at X_0, then for any u harmonic > 0 such that $\frac{u}{h}$ is bounded, $\frac{u}{h} \xrightarrow[\mathfrak{F}]{} \Lambda_u$.

c) If $f \geq 0$ on Δ is h-resolutive, $\frac{\mathcal{D}_{f,h}}{h}$ has a fine limit at X_0, that may be written $\int f \, d\nu^{X_0}$ (ν independant of f); measurability ν = measurability μ_h and the same sets of measure zero.

A similar study cannot go so far for a general compact boundary (§ 1), but results on the behaviour may be deduced from the important Naïm's theorem, that when $\mathcal{a}_h$ is satisfied, the general Dirichlet problem of § 1 is equivalent to a corresponding Dirichlet problem for the Martin boundary (thanks to an application of the first boundary on Δ_1, upto two sets respectively, h-negligible). Δ appears as the most important compact boundary, as well as the Euclidean one, and a deep comparison is useful, although not enough developed till now; however, the final basic topology is the fine one.

Important applications and possible extensions are obvious to functions of a complex variable, to Riemann surfaces (which are Green spaces when hyperbolic) to the correspondance between two such hyperbolic surfaces and their Martin boundaries. See Brelot [17], [18], [29], Naim [1], Constantinescu-Cornea [1] Doob [5], [7].

The next chapter will deepen the fine topology and study its connections with old classical results.

CHAPTER XVII

Comparison of both Thinnesses.

Fine Limits and Non-Tangential Limits.

(Classical Case. Examples).

1. Some examples of comparison.

Theorem XVII, 1. *Let us consider a Green space Ω and a polar point $x_0 \in \Omega$. For any set of $\Omega_1 = \Omega \setminus \{x_0\}$, the thinness at x_0 in Ω is equivalent to the minimal thinness relative to the minimal function $G^{\Omega}_{x_0}$.*

Proof. We have seen (Theorem XIV, 6) that $G^{\Omega}_{x_0}$ is minimal on Ω_1.

Now $e \subset \Omega_1$ is thin at x_0 in Ω iff $(R^e_{G_{x_0}^{\Omega}})_{\Omega} \not\equiv G^{\Omega}_{x_0}$. But

$(R^e_{G_{x_0}})_{\Omega} = (R^e_{G_{x_0}})_{\Omega_1}$ on Ω_1, because the positive superharmonic functions on Ω_1 are the restrictions of the positive superharmonic functions on Ω. Therefore, the thinness at x_0 is equivalent to the minimal thinness for $G^{\Omega}_{x_0}$. Another form of the proof comes from the identity of potentials in $\Omega \setminus \{x_0\}$ and potentials in Ω of a measure not charging $\{x_0\}$.

Exercise. If x_0 is non-polar in Ω (with dim $\geqslant 3$) the thinness at x_0 implies the minimal thinness for the minimal function $G^{\Omega}_{x_0}$ of Ω_1, but the converse is not true.

Actually, the minimal thinness is equivalent to the thinness of the set given by an inversion of the image of a neighbourhood of x_0. (See Brelot [6]).

An extension of the theorem will need two lemmas:

Lemma XVII, 2. Consider a domain ω in a Green space Ω, an irregular boundary point $x_0 \in \Omega$ of ω; $U = G^{\Omega}_{x_0} - R^{\complement\omega}_{G^{\Omega}_{x_0}}$ is in ω a minimal positive harmonic function.

Proof. In fact, consider in ω, u harmonic $\geqslant 0$, $\leqslant U$; if u_1 is the continuation on Ω by $G^{\Omega}_{x_0}$ of $u + R^{\complement\omega}_{G^{\Omega}_{x_0}}$ on ω, $\hat{u}_1$ is a potential; because if v is a potential, infinite on the set $e \subset \partial\omega \cap \Omega$ where $\complement\omega$ is thin, $u_1 + \frac{1}{n} v$ is superharmonic ≥ 0 and $\widehat{\lim (u_1 + \frac{1}{n} v)}$ is superharmonic, equal to u_1 outside e, therefore equal to $\hat{u}_1$ and is locally bounded outside x_0; as $u_1 \leq G^{\Omega}_{x_0}$, $\hat{u}_1$ is a potential, and the associated measure μ does not charge e except perhaps $\{x_0\}$[51]. Hence

$\hat{u}_1 = V + \alpha G_{x_0}$, V being the potential of the restriction μ' of μ on $\complement\{x_0\}$.

$$\hat{R}^{\complement\omega}_{\hat{u}_1} = \hat{R}^{\complement\omega}_{V} + \alpha \hat{R}^{\complement\omega}_{G^{\Omega}_{x_0}}, \alpha \geq 0.$$

But $\hat{R}^{\complement\omega}_{\hat{u}_1} = R^{\complement\omega}_{G^{\Omega}_{x_0}}$ because $\hat{u}_1 = G^{\Omega}_{x_0}$ q.e. on $\complement\omega$.

$\hat{R}^{\complement\omega}_{V} = V$ because μ' is supported by the base of $\complement\omega$.

Hence $u = \hat{u}_1 - \hat{R}^{\complement\omega}_{G^{\Omega}_{x_0}} = \alpha(G^{\Omega}_{x_0} - R^{\complement\omega}_{G^{\Omega}_{x_0}}) = \alpha U$ on ω.

Lemma XVII, 3 (Naïm). As a continuation of the previous lemma, consider a set $e \subset \omega$;

$$(\hat{R}^{e \cup \complement\omega}_{G^{\Omega}_{x_0}})_{\Omega} = (\hat{R}^{e}_{U})_{\omega} + (\hat{R}^{\complement\omega}_{G^{\Omega}_{x_0}})_{\Omega} \text{ on } \omega.$$

Proof. For any superharmonic function $u \geq 0$ on Ω, majorizing $G^{\Omega}_{x_0}$ on $e \cup \complement\omega$, $u - R^{\complement\omega}_{G^{\Omega}_{x_0}}$ is ≥ 0 and majorizes U on e, therefore $(R^{e}_{U})_{\omega}$ on ω. Hence the first member majorizes the second one. Conversely, $(\hat{R}^{e}_{U})_{\omega} + \hat{R}^{\complement\omega}_{G^{\Omega}_{x_0}}$ on ω, continued by $G^{\Omega}_{x_0}$ on $\complement\omega$, is a function U_1 such that $\hat{U}_1$ is a superharmonic ≥ 0 (even a potential) on Ω. Same proof as at the beginning of the previous proof. But $\hat{U}_1$ majorizes $G^{\Omega}_{x_0}$ q.e on $e \cup \complement\omega$, hence majorizes $(\hat{R}^{e \cup \complement\omega}_{G^{\Omega}_{x_0}})_{\Omega}$. Hence U_1 too.

Theorem XVII, 4 (Naïm [1]). Consider in a Green space Ω, a domain ω, with an irregular boundary point $x_0 \in \Omega$ (which implies x_0 is polar). The thinness of $e \subset \omega$ at x_0 in Ω is equivalent to the minimal thinness relative to the minimal function $G^{\Omega}_{x_0} - R^{\complement\omega}_{G^{\Omega}_{x_0}}$ on ω, i.e. at the corresponding point X_0 of the Martin boundary.

In other words, the correspondance of the spaces $\omega \cup \{x_0\}$ and $\omega \cup \{X_0\}$

(51) If w is superharmonic ≥ 0 on ω with associated measure μ and finite on a polar set e, then the inner μ measure of e is zero.

with the fine and minimal fine topologies, defined by identity on ω and $x_o \Longleftrightarrow X_o$, is a homeomorphism.

Proof. If e is thin, as $\complement\omega$ is thin, $e \cup \complement\omega$ is thin at x_o; $(\hat{R}^{e\cup\complement\omega}_{G^{\Omega}_{x_o}})_{\Omega} \not\equiv G^{\Omega}_{x_o}$. If there was equality on ω, there would be equality q.e on $\omega \cup \complement\omega = \Omega$, therefore everywhere. There exists x_1 where $(\hat{R}^e_U)_\omega + R^{\complement\omega}_{G^{\Omega}_{x_o}} \neq G^{\Omega}_{x_o}$, or $(\hat{R}^e_U)_\omega\ (x_1) \neq G^{\Omega}_{x_o}(x_1) - R^{\complement\omega}_{G^{\Omega}_{x_o}}(x_1) = U(x_1)$. This means the minimal thinness. Conversely, the minimal thinness implies $(\hat{R}^e_U)_\omega \neq U$ at some $x_2 \in \omega$. Therefore, $\hat{R}^{e\cup\complement\omega}_{G^{\Omega}_{x_o}}\ (x_2) \neq G^{\Omega}_{x_o}(x_2)$, i.e. the thinness of $e \cup \complement\omega$, hence the thinness of e.

Exercise. Suppose e is thin at x_o; $\hat{R}^{\complement\omega}_{\hat{R}^e_{G^{\Omega}_{x_o}}}$ is the greatest harmonic minorant of $\hat{R}^e_{G^{\Omega}_{x_o}}$ on ω, therefore, $\hat{R}^e_{G^{\Omega}_{x_o}} - \hat{R}^{\complement\omega}_{\hat{R}^e_{G^{\Omega}_{x_o}}}$ is a potential on ω. This implies the minimal thinness (no use of lemma 3).

2. Examples in a half-space ω of $\mathbb{R}^n$. (See Chiefly M[me] Lelong [1]).

The euclidean closure of ω in the Alexandroff compactification of $\mathbb{R}^n$ is also the Martin space $\hat{\omega}$.

Theorem XVII, 5. In a half space ω of $\mathbb{R}^n$, $n \geq 3$, consider a normal Stolz domain $\mathscr{C} \neq \omega$, i.e. an open cone of vertex x_o on $\partial\omega \cap \mathbb{R}^n$ with axis of revolution orthogonal to the hyperplane $\partial\omega$ and limited to its intersection with a ball of centre x_o. The thinness of $e \subset \mathscr{C}$ at x_o in $\mathbb{R}^n$ is equivalent to the minimal thinness at x_o in ω.

Proof. We introduce the sets $I_p = \{s^{p+1} < |x-x_o| \leq s^p\}$, $0<s<1$, $I^e_p = I_p \cap \mathscr{C}$. A criterion of thinness of e is $\sum_p \gamma_p / s^{p(n-2)} < +\infty$. ($\gamma_p$ outer capacity of $e_p = e \cap I_p$ in $\mathbb{R}^n$). It is another form of the general Wiener criterion (th. IX, 10). For real φ, ψ depending on precised conditions (for instance on variables like p in given variable fields), let us say they are comparable and let us write $\varphi \simeq \psi$ if $\frac{\varphi}{\psi}$ (when it has a sense) lies between two fixed members > 0 (depending only on the considered conditions). Now, an elementary calculus in $\mathbb{R}^n (n \geq 3)$ shows that the outer Greenian capacity γ^ω_p of e_p in ω

is comparable to γ_p (variable p). That comes from the fact that

$$\frac{1}{|x-y|^{n-2}} \Big/ G^{\omega}(x,y) \simeq 1$$ for x and y variable on $I_p \cap \mathcal{E}$ (p variable). Suppose the thinness. Then the series $\left\{\gamma_p^{\omega} / s^{p(n-2)}\right\}$ converges. Now γ_p^{ω} is the total measure corresponding to $(\hat{R}_1^{e_p})_{\omega}$. For $x \in \mathcal{E}$, $G^{\omega}_{y_0}(x) \simeq \delta_x$ (distance of the variable point x to $\partial\omega$) $\simeq |x-x_0|$. This is on $I_p \cap \mathcal{E}$ comparable to s^p (variable p).

Therefore, $(R_1^{e_p})_{\omega}(y_0) = \gamma_p^{\omega} s^p$ and

$$\gamma_p^{\omega} s^{-p(n-2)} \simeq (\hat{R}_1^{e_p})_{\omega}(y_0)\, s^{-p(n-1)} \quad \text{(variable p).}$$

But on I_p, $K_{x_0}(x) \simeq \dfrac{\delta_x}{|x-x_0|^n} \simeq s^{-p(n-1)}$ (variable p).

Hence $\left\{\hat{R}^{e_p}_{K_{x_o}}(y_0)\right\}$ converges, $\sum_N^{\infty} \hat{R}^{e_p}_{K_{x_o}}(y_0) \to 0$ as $N \to \infty$ and

$R_{K_{x_0}}^{e \cap \{|x-x_0|<r\}}(y_0) \to 0$ as $r \to 0$. We conclude e is minimal thin.

The converse is obviously a consequence of the convergence of $\left\{\hat{R}^{e_p}_{K_{x_o}}(y_0)\right\}$ which means the minimal thinness according to a Wiener type criterion given in the following exercise.

Complements. According to Mme Lelong [1], the equivalence given in the theorem does no more hold(52) for any $e \subset \omega$, but the thinness still implies the minimal one.

The similar study in $\mathbb{R}^2$ is more difficult; the last implication still holds for any $e \subset \omega$ (53) (Jackson [1]), but not the converse.

In other words, the identical mapping of the space $\omega \cup \{x_0\}$ of $\mathbb{R}^n$ ($n \geq 2$) with minimal fine topology on the same set with fine topology is continuous, but not the inverse mapping.

Further developments are to be seen in Mme Lelong [1], Brelot-Doob [1], Naïm [1], Jackson [1]

---

(52) There exist minimal thin sets which are not thin (even for $n \geq 2$), because for example, of the unthinness of the closure of the following set: a ball minus an interior ball tangent at x_0.

(53) This is a rectification of the independance of both thinnessindicated without proof by Mme Lelong.

Exercises. a) A criterion of minimal thinness of e in a half-space ω of $\mathbb{R}^n$ (even for $n \geqslant 2$) is, with the same definition of I_p, the convergence of the series $\left\{ R_{K_{x_0}}^{e \cap I_p}(y_0) \right\}$ (Mme Lelong). In $\mathbb{R}^2$, for e in a Stolz domain, another obvious equivalent criterion is the convergence of the series $\left\{ \gamma_p^{\omega} \right\}$.

See some examples and applications in Brelot-Doob [1].

b) Again for $n \geqslant 2$, with the same notations, the condition $R_{K_{x_0}}^{e \cap I_p}(y_0) \longrightarrow 0$ (independant of s and y_0) may define the minimal <u>semi-thinness</u>.

A set e satisfying $\dfrac{\delta_x}{|x - x_0|} \longrightarrow 0$ $(x \in e,\ x \longrightarrow x_0)$ is minimal semi-thin.

c) In $\mathbb{R}^n$ $(n \geqslant 2)$ the semi-thinness of e (see Chap IX, § 6) implies the minimal semi-thinness in ω.

That may be deduced of the same property for e in a Stolz domain. In this case, use for $n \geqslant 3$, $\gamma_p \simeq \gamma_p^{\omega}$ and conclude that $\gamma_p / s^{p(n-2)} \simeq (R_{K_{x_0}}^{e \cap I_p}(y_0))_{\omega}$. For $n = 2$, the semi-thinness means $n\gamma_p \longrightarrow 0$ and the minimal one (in a Stolz domain) to $\gamma_p^{\omega} \longrightarrow 0$. According to an argument of Jackson, the first condition implies the second one.

d) In any Green space Ω, the semi-thinness of e at $X \in \Delta_1$ may be defined in the following ways. Defining $\sigma_\lambda = \left\{ x \,\middle|\, \dfrac{K_X(x)}{G_{y_0}^{\omega}(x)} \geqslant \lambda \right\}$,

α) Condition $R_{\lambda G_{y_0}^{\omega}}^{e \cap \sigma_\lambda} \longrightarrow 0$ $(\lambda \longrightarrow \infty)$ (Brelot-Doob)

β) Condition $R_{K_X}^{\sigma_{t^p} - \sigma_{t^{p+1}}} \longrightarrow 0$ $(p \longrightarrow \infty)$, $t > 1$ (Doob, unpublished).

Compare with thinness and show the equivalence with the previous definition (b) in a half-space of $\mathbb{R}^n$.

It is easy to improve the previous examples by changing ω in its intersection with a ball of centre x_0. But it would be desirable to generalize these implications at least for a regular domain whose euclidean closure is also the Martin one.

3. Comparison of statistical thinness (Brelot [26], [27].

As such implications are not known for general domains, we shall consider weaker conditions and implications that will remain valid in axiomatic theories.

Definition XVII, 5. Consider a domain ω in a Green space Ω and a boundary point x_0 of ω in Ω; X minimal of the Martin boundary of ω (i.e. $X \in \Delta_1$) is said to be associated to x_0 if for any neighbourhood δ of x_0, $\Omega \setminus \delta$ is (minimal) thin at X, i.e. if $(\delta \cap \omega) \cup \{X\}$ is a fine neighbourhood of X (See another equivalence in Naim [1]).

Theorem XVII, 6. If $e \subset \omega$ is 1-statistically thin in Ω, on $\alpha \subset \partial\omega \cap \Omega$ (in particular thin at every point of α), it is 1-statistically minimal thin on the set E of points of Δ_1^{ω} associated to the points of α.

Proof. If δ is a variable neighbourhood of α, $\{\delta \cap e\}$ is vanishing in Ω (Th. VIII, 16), therefore also in ω; but $(\delta \cap \omega) \cup E$ is a fine neighbourhood of E and the intersection with e is $\delta \cap e$. Therefore, if $\mathcal{V}$ is a variable fine neighbourhood of E in $\hat{\hat{\omega}}$ (i.e. $\omega \cup \Delta_1^{\omega}$), $\{\mathcal{V} \cap e\}$ is vanishing and that implies e is minimal thin μ_1- everywhere on E (Th. XV, 15).

Extension. Suppose W is superharmonic > 0 in Ω and H, a minorant harmonic > 0 in ω.

If e is W-statistically thin on $\alpha \subset \partial\omega \cap \Omega$, it is H-statistically minimal thin on E (set of points of Δ_1^{ω} associated to the points of α).

Same argument with the extensions of Chap XV, § 5.

4. Angular (i.e. non-tangential) limits and fine limits for harmonic functions in a half-space $\omega \subset \mathbb{R}^n$.

Consider a function f (real for example) in a half space ω and a boundary point x_0; λ is commonly called non-tangential or angular limit of f at x_0 if $f \to \lambda$ (in the euclidean topology) on every Stolz domain as defined earlier (vertex x_0); λ is called an angular limit value (an adherent value) if $f(x_n) \to \lambda$ for a sequence $x_n \to x_0$, x_n belonging to a Stolz domain.

We already considered the sets of ω, "totally tangent" at x_0, i.e. such that $\dfrac{\delta_x}{|x-x_0|} \to 0$, $x \in e$, $x \to x_0$. The complementary sets in ω form a filter $\mathcal{J}^{a}_{x_0}$. Now, the previous angular notions (limit, limit value) are equivalently limit or limit value according to $\mathcal{J}^{a}_{x_0}$ (remark of Doob).

We will consider also the filters $\mathcal{J}^{m}_{x_0}$, $\mathcal{J}^{s}_{x_0}$ of the complementary sets in ω of (minimal) thin, semi-thin sets, respectively, at x_0. To these filters correspond notions of (minimal) fine limits, as we know, and semi-fine limits. We know that the first ones are equivalent to the euclidean limits outside of a suitable thin set and one may prove that it is the same for semi-fine limits and semi-thin sets.

Note that $\mathcal{J}^{s}$ is finer than $\mathcal{J}^{m}$ and than $\mathcal{J}^{a}$ (see § 2 exericse (b)).

Lemma XVII, 7. In the half-space ω, let us consider a variable ball B_x (center x, radius R_x, with $R_x = \alpha\delta_x$, $0 < \alpha$ fixed < 1. Then, for fixed points x_0, y_0 of $\mathbb{R}^n$, $x_0 \in \partial\omega$, $y_0 \in \omega$,

$$\liminf_{\substack{x \to x_0 \\ x \in \omega}} R_{\tau}^{B_x}(y_0)/\delta_x^{n-1} > 0.$$

Proof. This is implied by the case where $x_0 - y_0$ is orthogonal to the hyperplane $\partial\omega$ and x on this normal. Then, if ω_x is the half-space $\{\delta_x > \alpha\} \subset \omega$, $R_1^{B_x}(y_0)$ majorizes the harmonic measure in ω_x at y_0 of $\partial\omega_x \cap B_x$ which is upto a factor, equivalent to R_x^{n-1} or δ_x^{n-1}.

Classical results of Fatou type on angular or normal boundary limits a.e for harmonic or superharmonic functions (Calderon, Stein, Zygmund...) can be deduced now from the following general result on fine limits, they have inspired.

Theorem XVII, 8 (Brelot-Doob [1], Constantinescu-Cornea [1]). For two strictly positive harmonic functions u and h in the half-space ω, (or only in a neighbourhood of a boundary point $x_0 \in \mathbb{R}^n$), any angular limit value of $\frac{u}{h}$ at x_0 is a (minimal) fine, even a semi-fine limit value. Therefore, fine limit implies equal angular limit. There is equivalence between angular limit and semi-fine limit.

Proof. Consider $x_p \to x_0$, x_p on a Stolz domain $\mathcal{C}_1 \subset \omega$, such that $\frac{u(x_p)}{h(x_p)} \to \lambda$ (for example finite; if not, a similar argument). Given $\varepsilon > 0$, we will define a series of balls B_p of center x_p, in another larger Stolz domain $\mathcal{C}_2$ and on which $\left| \frac{u(x_p)}{h(x_p)} - \lambda \right| < \varepsilon$ ($p >$ a suitable p_0) and prove that their union is not thin, even not semi-thin ; therefore, any set of $\mathcal{T}^m$ or $\mathcal{T}^s$ will contain a point where $\left| \frac{u}{h} - \lambda \right| < \varepsilon$, and that proves that any angular limit value is a semi-fine one.

Recall that, if w is any harmonic function > 0 in a ball (center x_0, radius ρ)

$$\sup_w \left(\sup_x \frac{w(x)}{w(x_0)} \right)$$

for x varying in a concentric ball (radius $\alpha\rho$) is a function $\theta(\alpha) \underset{\alpha \to 0}{\to} 1$ independent of ρ (consequence of classical Harnack inequalities).

Now considering our sequence x_p, we may choose first a p_0 such that $p > p_0$ implies

$$\lambda - \varepsilon/2 < \frac{u(x_p)}{h(x_p)} < \lambda + \varepsilon/2$$

then in $\mathcal{C}_2$ a ball B'_p (center x_p, radius $\rho_p = \beta\delta_{x_p}$, with a fixed small enough β). Now we shall have

$$\frac{\lambda - \varepsilon/2}{\theta^2(\alpha)} < \frac{u}{h} < (\lambda + \varepsilon/2)\,\theta^2(\alpha)$$

in the concentric ball B''_p of radius $\alpha\rho_p$ ($\alpha < 1$) or $\alpha\beta\delta_{x_p}$.

For a suitable α (independent of p, like β)

$$\lambda - \varepsilon < \frac{u}{h} < \lambda + \varepsilon \quad \text{in every } B''_p \ (p > p_0) .$$

According to the previous lemma,

$$R_1^{B''_p}(y_0) > K\delta^{n-1}_{x_p} \quad (K \text{ independent of } p) .$$

As $K_{x_0}(x) \simeq \delta^{n-1}_{x_p}$ on B''_p,

$$\inf R_{K_{x_0}}^{B''_p}(y_0) > \eta > 0 .$$

Any B''_p intersects a finite number ν_p of I_q such that $\nu_p \leq \nu$ independant of p (by using an argument of homothety). For one intersection i,

$$R^{i}_{K_{x_o}}(y_0) > \frac{\eta}{\gamma} .$$

There exist arbitrarily large q for which

$$R^{\bigcup_{p > p_0} B''_p \cap I_q}_{K_{x_o}}(y_0) > \frac{\eta}{\gamma} .$$

therefore, this function of q does not tend to zero as $q \to \infty$; therefore $\bigcup B_p$ is not thin and even is not semi-thin. These B''_p are the wanted B_p .

So, semi-fine limit of $\frac{u}{h}$ implies angular limit. The converse is obvious because $\mathcal{T}^s$ is finer than $\mathcal{T}^a$.

Complements. Indications (See Brelot-Doob [1]). Instead of $u > 0$, $h > 0$, it is sufficient to suppose $h > 0$, $\frac{u}{h}$ bounded from a side (in a neighbourhood of x_0) and in $\mathbb{R}^2$ even no restriction on u (harmonic) as a consequence of special properties of thinness in $\mathbb{R}^2$. Propositions may be adapted for functions defined close to x_0 , only in an open cone (general Stolz domain (54), vertex x_0).

It is important to note the counter example of Choquet, that a positive harmonic function in a half-plane may have an angular limit at x_0 , but no fine limit.

Converse, but statistical properties are also studied in Brelot-Doob's paper. For example, angular limit for any function (taking its values even in a compact metric space) at all points of a set $e \subset \partial\omega$ implies a. e. (Lebesgue meaning) on e , an equal (minimal) fine limit.

It would be interesting to compare also fine limit and angular limit for BLD harmonic functions.

Exercises. 1) Consider still in $\mathbb{R}^n$ $(n \geq 2)$, a half-space ω , a Stolz domain $\mathcal{E}$ with vertex $x_0 \in \partial\omega$ and $e \subset \omega$, minimal thin or only semi-thin at x_0 . Let us introduce $\mathcal{E}'$ smaller, I'_p intersection of $\mathcal{E}'$ with

(54) i. e. an open conic domain of revolution, contained with its closure (minus x_0) in the half-space and limited to its intersection with a ball, center x_0 .

$\{x \mid \theta_1 s^{p+1} < |x-x_0| < \theta_2 s^p\}$, $0 < \theta_2 < 1 < \theta_1$. The outer harmonic measure of $\partial I_p \cap e$ for I_p on I'_p tends to zero uniformly (as $p \to \infty$).
Hint: (Direct application of the definitions and of the comparison of the previous harmonic measure with $R_1^{\partial I_p \cap e}$).

As an application, we may give another proof of the following part of the previous theorem:

2) If $\frac{u}{h}$ has a fine (or semi-fine) limit λ at x_0 (u,h harmonic > 0 on ω), it has an equal angular limit.

Hint: If $\lambda = +\infty$, we may see that on I'_p, $\frac{u}{h}$ tends to $+\infty$; if λ is finite, u/h remains bounded in any Stolz domain and tends to λ outside of an open semi-thin set e. We introduce $u' = u$ outside e, $= \lambda h$ on e, and $u - H_{u'}^{I'_p}$. Using (1), we see the quotient by h tends to zero on I'_p.

Hence the wanted result and as previously the equivalence of angular and semi-fine limit.

Application to the old Fatou theorem. Comparing th XVI, 16 and th. XVII, γ in a half space ω, we get with u and h harmonic > 0 on ω, the existence for $\frac{u}{h}$ of an angular limit μ_h - a.e on the hyperplane $\partial\omega$ (if $h = 1$ as in the original Fatou case, μ_1 - a.e means a.e according to the Lebesgue measure on $\partial\omega$). Similar expression in a ball, as Fatou did.

An improvement of Calderon-Carleson has been again improved by Doob (See Brelot-Doob [1]) as we will only indicate: h harmonic > 0 given in the half-space, but u harmonic given in ω_0 open $\subset \omega$; ω_0 contains a variable general Stolz domain whose vertex describes a boundary set e of $\partial\omega$. In each such domain, $\frac{u}{h}$ is supposed to be bounded from a side above or below, depending on the vertex. Then $\frac{u}{h}$ has a finite angular limit on e, except on a set of Lebesgue measure zero and on a set of μ^h-measure zero.

Important Remark. There is no similar property for superharmonic functions. See a counter example of Zygmund ($h = 1$, in $\mathbb{R}^2$), in Tolsted [1]. This shows the advantage and the power of the fine limit.

5. Comparison of angular, fine and normal limits (Short indications).

Doob [7] succeeded also to deduce from his general result on fine limits, the classical results on normal limits and to improve and develop this question.

Consider first a set A in a half space ω of $\mathcal{R}^n$. A normal limit point x_0 of A is a point of $\partial\omega \cap \mathcal{R}^n$ on the euclidean closure of $A \cap n_{x_0}$, where n_{x_0} is the normal at $\partial\omega$ in x_0. For almost (Lebesgue measure on $\partial\omega$) such a point x_0, A is (minimal) unthin at x_0 (i.e. x_0 is a (minimal) fine limit point)[55].

Consider now a function f from ω to a compact metric space E and denote N_{x_0}, A_{x_0}, F_{x_0} the cluster sets of f relative to the approach of x_0 on n_{x_0}, or according to $\mathcal{F}^a_{x_0}$ or $\mathcal{F}^m_{x_0}$. Then, a.e on $\partial\omega$, $N_{x_0} \subset \mathcal{F}^m_{x_0}$. As a consequence, fine limit $\Rightarrow$ normal (equal) limit a.e

When f is superharmonic, we get more: at every boundary point x_0 of the hyperplane except on a set of Lebesgue measure zero (on $\partial\omega$) with $E = [-\infty, +\infty]$

α) $-\infty \in F_{x_0} \subset A_{x_0}$

or β) $\exists$ a finite fine limit, with equal normal limit, but with an angular limit or not.

If f superharmonic has a harmonic minorant $u \leq 0$, this u possesses finite fine limits a.e on $\partial\omega$ and (α) is excluded; therefore, f has a normal limit a.e. This is an equivalent form of a classical result of Littlewood-Privaloff.

See Doob for further details and proof. Let us only add that as an application to function theory, Doob [7] was led to an improvement of the classical Plessner theorem. Given f meromorphic in the half-plane ω, Plessner proved that a.e

α) A_{x_0} is the extended plane or

---

(55) The proof of Doob is complicated. The result implies the similar property with the ordinary thinness (on $\mathcal{R}^n$). Note that this latter property (even extended to any $e \subset \mathcal{R}^n$) is easier and the direct proof gives even "(q.e)" on $\partial\omega$, instead of (a.e) on $\partial\omega$.

β) A_{x_0} is one point (angular limit)[56].

Now, except on another set of measure zero on $\partial\omega$, we have more: (α) is divided in two cases:

F_{x_0} is the extended plane too or F_{x_0} and N_{x_0} are one and the same point.

(β) is completed by: F_{x_0} is also one point (and the same as A_{x_0}). The part without normal limit was also given by Constantinescu-Cornea.

---

(56) Note that (even for f taking its values in a general previous E), A_{x_0} is a.e also the cluster set defined in every Stolz domain.

CHAPTER XVIII

Martin Space and Minimal Thinness

In Axiomatic Theories - Short Survey

1. Various hypotheses and notations

We start first from the Brelot-axiomatic recalled in Chap XI. Same space Ω, but with a countable base (although this is not always necessary). Axioms 1,2,3 and the existence of a potential > 0 are supposed. (A_1) will denote these hypotheses. We will complete Chap XI, apply Chap XII and extend Chap XIV, § 5. We choose on the set S of the differences of positive superharmonic functions, the Hervé's T-topology and a compact metrizable base B of S^+. Δ_1 will denote the set of the minimal harmonic functions on Ω belonging to B, i.e. of the extreme elements of B (identifiable with the minimal boundary of Ω)

(A_1^P) will denote A_1 with the hypothesis of proportionality (of all potentials of the same point support).

Then the potentials of this kind in B, denoted p_x (support $\{x\}$) form a set which is homeomorphic to Ω (Gowrisankaran [1]). Its T-closure is the generalized Martin space $\hat{\Omega}$ (so called, because it is in the classical case homeomorphic to the previous Martin space). The Martin boundary $\Delta = \hat{\Omega} - \Omega$ contains Δ_1 whose points X are denoted also sometimes $p_X(y)$ as harmonic functions on Ω. In the Riesz integral representation of a superharmonic function v, the corresponding measure μ_v on B may be considered as a measure on $\hat{\Omega}$, supported by $\Omega \cup \Delta_1$. Then $v(y) = \int_{\Omega} p_x(y)\, d\mu_v + \int_{\Delta_1} p_X(y)\, d\mu_v$.

(A_2) The lacking of some $p_x(y)\lambda(x)$ analogous to the symmetric Green function will impose the use of the Hervé's adjoint sheaf, therefore, leads to the supplementary hypothesis of a base of "completely determining domains" (see Chap XI § 5), and that gives with A_1^P the rich set A_2.

A suitable example of B is given by the condition $\int v \, d\rho_{x_0}^{\omega_0} = 1$ ($v \in S^+$, ω_0 a completely determining domain, $x_0 \in \omega_0$).

2. Fine topology on $\check{\Omega} = \Omega \cup \Delta_1$ (See Brelot [33])

We will continue to extend Chap XIV by using Chap XII.

With (A_1) above, the thinness of $e \subset \Omega$ at $X \in \Delta_1$ is defined by $R_X^e \neq X$, or $\hat{R}_X^e$ is a potential. The complementary sets of these thin sets on Ω form as we know, a filter $\mathcal{F}_X$ according to which notions of limit are said minimal fine. Let us give a topological interpretation, more precise than Chap XII.

Let us say the topology θ' on $\check{\Omega}$ is a minimal continuation of the topology θ on Ω, if it induces θ on Ω and if the neighbourhoods of any $X \in \Delta_1$ intersect Ω according to the sets of $\mathcal{F}_X$. Now, there exists for the fine topology on Ω, minimal continuations on $\check{\Omega}$; they make Ω open. There is a finest one (inducing on Δ_1, the discrete topology) and a coarsest one. In case of A_1^P, there is only one such continuation.

Inner interpretation (with A_2). Consider on Ω the family Φ_1 of the adjoint hyperharmonic functions $\geqslant 0$, the subfamily Φ_2 of the adjoint potentials (and $+\infty$), and the subfamily Φ_3 of the particular adjoint potentials of the form $\int p_X(y)\, d\mu(y)$ ($\mu \geqslant 0$ on Ω). Denote Φ_1', Φ_2', Φ_3', the family of functions on $\check{\Omega}$, we get by lower-semi-continuity according to the T-topology (Martin topology) on $\check{\Omega}$. Denote T_1, T_2, T_3 and T_1', T_2', T_3' the topologies on Ω (resp $\check{\Omega}$) with initial topology T, associated to the previous families. Then T_1', T_2', T_3' are the unique minimal continuations of T_1, T_2, T_3.

Note that T_1, T_2 are the adjoint fine topologies of Ω, whose minimal continuation is interpreted as an inner topology T_1' or T_2' on $\check{\Omega}$.

Note also the partial result that any adjoint superharmonic function $\geqslant 0$ on Ω has at any $X \in \Delta_1$, a $\lim_{\mathcal{F}_X}$ equal to its $\liminf_T$.

Moreover, the inner thinness or unthinness at $X \in \Delta_1$ relative to Φ_1', Φ_2' or Φ_3' on $\check{\Omega}$ with initial T-topology are strong.

See more details and proofs, criterion for $X \in \Delta$ to be minimal, in Brelot [33].

3. Dirichlet problem and boundary behaviour. (Extension of Chap XV, XVI).

With A_1 and first also with axiom D, Gowrisankaran [1] extends the key-theorem (criterion for $R_h^e = h$, harmonic > 0) and proves that $\frac{v}{h}$ (v, a potential > 0) has a fine limit 0, μ_h a.e on Δ_1; he studies a Dirichlet problem for h-harmonic functions on Ω with the minimal fine topology for the boundary conditions. Supposing a condition R_h (any uniform T-continuous function on Δ_1 is h-resolutive), he develops the previous general Dirichlet problem, givesthe resolutivity-theorem and deducesthe Doob type result for functions $\frac{u}{h}$ (u superharmonic > 0) i.e., the existence of a fine limit μ_h - a.e on Δ_1. In case of A_1^P, R_h is always satisfied and the Dirichlet problem for h-harmonic functions and T-topology on $\hat{\Omega}$ may be developed; the cases of resolutivity and solutions are the same in both problems. Later Gowrisankaran [4] succeeded to develop independantly another Dirichlet problem, like in theorem XVI, 11, with A_1 alone, <u>without</u> D without any R_h-type condition as follows:

key-criterion of the property $R_h^e = h$, but only for e open, boundary property of any potential w ($\frac{w}{h}$ has fine limit 0, μ_h- a.e), Dirichlet problem with boundary condition

$$\text{fine lim inf } \frac{v}{h} \geqslant \text{ given } f \text{ on } \Delta_1, \ \mu_h \text{ a.e}$$

$$\text{or fine lim sup } \geqslant f \qquad \mu_h \text{ a.e.}$$

The envelopes (inf of such hyperharmonic v, lower bounded) are the same for both problems, as well as the resolutivity (equivalent to the μ_h-summability of f) and the solutions. Hence, with A_1 alone, the extension of the Doob's property (finite fine limit of $\frac{v}{h}$, μ_h - a.e. for any superharmonic $v \geqslant 0$).

Gowrisankaran [5] studied also without D, a Dirichlet problem for any compactification of Ω, extending Chap XVI, § 1 and compared with the previous solutions. Other researches on compactification and Dirichlet problem were developed by Constantinescu-Cornea [3] and Loeb, B. Walsh.

4. Let us mention also some other extensions of the classical case; with various hypotheses:

a) Fatou-Doob type boundary property in the case of a lower bounded only on a fine (minimal) neighbourhood of every point of a μ_h-measurable set $E \subset \Delta_1$

(with A_1, D).

b) use of uniform integrability(for characterization of the solutions of a Dirichlet problem).

c) behaviour of various capacities of decreasing sets (for example finely closed) (with A_1, D and partly with A_2).

d) notion of W-polar sets on $\check{\Omega}$ (i.e. $e \cap \Omega$ is W-polar and $e \cap \Delta_1$ has μ_W-measure 0) with various characterizations (with A_1 or A_1^P).

e) Correspondance of two spaces with previous harmonic structures and correspondance of their minimal boundaries (Extension of Constantinescu-Cornea-Doob results and also of a Riesz theorem on 2 Riemann surfaces). Chiefly with A_1^P and D on both spaces.

f) Dirichlet problem for compact sets, notion of stable boundary point, and their use for a quasi-analyticity property in case of A_2.

g) Doubly harmonic functions u(x,y) (x,y on two spaces) and boundary behaviour.

h) Study of $\mathcal{H}^p$-harmonic functions, i.e. with a L^p-norm relative to the harmonic measure.

See for (a,g) Gowrisankaran [5,3] and J.B. Walsh [1]; for b,c,d Brelot [22, 30, 33], for e, Constantinescu-Cornea [3] and D. Sibony [1,2], for f, A. de la Pradelle [1,2] and for h, Lumer-Naim [3].

<u>Weaker axiomatics</u>. Minimal thinness and boundary may be considered in weaker axiomatics also, like the Bauer's one. Some results as the Doob's one hold good. Same extension for the correspondance of two spaces. D. Sibony [2] even extended the Doob's result in a larger frame, not necessarily of local character, containing also certain theories on excessive functions; he deduces a study of a Dirichlet problem for boundary conditions with minimal fine filters.

As a <u>final remark</u>, let us repeat there are other topological questions in potential theory, like the various uses of Hilbert spaces (See Deny [4]) and functional analysis (important researches of Loeb, B. Walsh, Mokobodzki [2] are most in the course of publication) and that we leave aside the wide field of the probabilistic interpretations. (See an introduction Bauer [7]).

Bibliography

L. V. Ahlfors and M. Heins, Questions of regularity connected with the Phragmen-Lindelöf principle (Annals of Maths. 50, n° 2, 1949 p.341 - 346).

H. Bauer, [1] Silovscher Rand und Dirichletsches Problem (Ann. Inst. Fourier 11, 1961, p. 89 - 136).

[2] Axiomatische Behandlung des Dirichletschen Problems für elliptische und parabolische Differential gleichungen (Math. Ann 146, 1962, p. 1 - 59).

[3] Weiterführung einer axiomatischen Potentialtheorie ohne Kern (Existenz von Potentialen) (Z. Wahrscheinlichkeitstheorie, 1, 1963, p. 197 - 229).

[4] Propriétés fines des fonctions hyperharmoniques dans une théorie axiomatique du potentiel (Colloque Paris-Orsay 1964 and Ann. Inst. Fourier 15/1, 1965, p. 137 - 154).

[5] Harmonische Räume und ihre Potentialtheorie (Lecture Notes 22, Springer 1966).

[6] Recent developments in axiomatic potential theory (Symposium of Loutraki, Lecture Notes 31, Springer 1967).

[7] Harmonic spaces and associated Markov Processes (Summer Course C.I.M.E. Stresa, 1969).

Blumenthal-R. Getoor, Markov Processes and Potential theory (Academic Press 1968).

N. Boboc, C. Constantinescu, A. Cornea, [1] Axiomatic theory of harmonic functions (Colloque Paris-Orsay and Ann. Inst. Fourier 15/1, 1965, p. 283-312.

[2] Axiomatic Theory of harmonic functions-Balayage (Ann. Inst. Fourier, 15/2, 1965, p. 37 - 70).

N. Boboc, A. Cornea, Behaviour of harmonic functions at a non-regular boundary point (Bull. Math de la Soc. Sc. Math. Phys. RSR 1965).

J.M. Bony, [1] Détermination des axiomatiques de theorie du potentiel dont les fonctions harmoniques sont differentiables. (Ann. Inst. Fourier 17/1, 1967, p.353-382).

[2] Opèrateurs elliptique, dègénèrés associés aux axiomatiques de la théorie du potentiel (Summer Course C.I.M.E. Stresa, 1969).

G. Bouligand, Fonctions harmoniques. Principes de Picard et de Dirichlet (Memoires des Sc. Math. XI, 1926).

M. Brelot, [1] Sur le potentiel et les suites de fonctions sousharmoniques (C.R. Ac. Sc. 207, 1938, p.1157).

[2] Critères de régularité et de Stabilité. (Bull. Ac. royale des Belgique 25, 1939, p.125 - 137).

M. Brelot, [3] Familles de Perron et problème de Dirichlet (Acta Szeged IX, 1939, p. 133 - 153).

[4] Points irréguliers et transformations continues en theorie du potentiel (J. de Math. 19, 1940, p. 319 - 337).

[5] Sur les ensembles effilés (Bull. Sc. Math. 68, 1944, p.12 - 36)

[6] Sur le rôle du point à l'infini dans la theorie des fonctions harmoniques (Ann. E.N.S. 61, 1944, p.301 - 332).

[7] Sur l'approximation et la convergence dans la théorie des fonctions harmoniques ou holomorphes (Bull. Soc. Math. France 73, 1945, p. 55 - 73).

[8] Minorantes sousharmoniques, extrémales et capacités (J. de Math. 24, 1945, p. 1 - 32).

[9] Le problème de Dirichlet ramifié (Ann. Univ. de Grenoble Math. Phys. 22, 1946, p. 167 - 200).

[10] Etude générale des fonctions harmoniques ou surharmoniques positives au voisinage d'un point frontière irrégulier (Ann. Univ. Grenoble, Math. Phys. 22, 1946, p.205 - 249).

[11] Quelques propriétés et applications du balayage (C.R. 227, 1948, p. 19).

[12] Sur le principe des singularites positives et la topologie de R.S. Martin (Ann. Univ. Grenoble Math. Phys. 23, 1948, p.119-138).

[13] Sur l'allure des fonctions harmoniques et sousharmoniques à la frontière (Math. Nach. 4, 1950, p. 298 - 307).

[14] La théorie moderne du potentiel (Ann. Inst. Fourier 4, 1952, p. 113 - 140).

[15] Etude et extension du principe de Dirichlet (Ann. Inst. Fourier 5, 1953 - 54, p. 371 - 419).

[16] On the behaviour of harmonic functions in the neighbourhood of an irregular boundary point (J. Anal. Math. 4, 1954 - 56, p.209 - 221).

[17] Le problème de Dirichlet. Axiomatique et frontière de Martin (J. de Math. 35, 1956, p. 297 - 335).

[18] Sur l'allure à la frontière des fonctions sousharmoniques ou holomorphes (Ann. Ac. Sc. Fenn. A. Math. 250/4, 1958).

[19] Axiomatique des fonctions harmoniques et surharmoniques dans un espace localement compact (Sem. Potentiel à Paris 2, 1958).

[20] Lectures on Potential Theory (Tata Institute n° 19, Bombay 1960, re-issued 1967).

M. Brelot, [20bis] Sur un théorème de prolongement fonctionnel de Keldych concernant le problème de Dirichlet (Journal Anal. Math VIII, 1960-61 p. 273 - 288).

[21] Introduction, axiomatique de l'effilement (Annali di Math. 57, 1962, p. 77 - 96).

[22] Intégrabilité uniforme, quelques applications à la théorie du potentiel (Sem. potentiel 6/1, 1962).

[23] Quelques propriétés et applications nouvelles de l'effilement (Sem. Pot. 6/1, 1962).

[24] On Martin boundary (Hiroshima University 1962; published in the Collection of Russian translations, Matematika 1965).

[25] Eléments de la théorie classique du potentiel (C.D.U. Paris, 1st edition, 1959, 4th edition 1969).

[26] Etude Comparée des deux types d'effilement (Colloque Paris-Orsay 1964 et Annales Inst. Fourier 15/1, 1965, p. 155 - 168).

[27] Aspect statistique et comparé des deux types d'effilement (Anais da Ac. Brasileira de ciências, 37, 1965).

[28] Axiomatique des fonctions harmoniques (Sem. Math. Sup res Montreal, Summer 1965).

[29] Théorie du potentiel et fonctions analytiques (Colloque de Erevan 1965).

[30] Capacity and balayage for decreasing sets (Symposium on probability and Statistics, Berkeley, Summer 1965).

[31] La topologie fine en théorie du potentiel (Colloque de Loutraki 1966, Lecture Notes n° 31, p. 36 - 47, Springer 1967).

[32] Recherches axiomatiques sur un théorème de Choquet concernant l'effilement (Nagoya Math. J. 30, 1967, p. 39 - 46).

[33] Recherches sur la topologie fine et ses applications (theorie du potentiel Ann. Inst. Fourier 17/2, 1967).

[34] Historical Introduction (Summer course C.I.M.E., Stresa 1969).

M. Brelot and G. Choquet, Espaces et lignes de Green (Ann. Inst. Fourier 3, 1951, p.199 - 263).

M. Brelot and J.L. Doob, Limites angulaires et limites fines (Ann. Inst. Fourier 13/2, 1963, p. 395 - 415).

A. Calderon, On the behaviour of the harmonic functions at the boundary (T.A.M.S, 68, 1950, p. 47 - 54).

H. Cartan, [1] Théorie du potentiel newtonien:énergie, capacité, suites de potentiels (Bull. Soc. Math. France 73, 1945, p. 74 -).

[2] Théorie générale du balayage en potentiel newtonien (Ann. Univ. Grenoble, Math. Phys. 22, 1946, p.221 - 280).

G. Choquet, [1] Theory of capacities (Ann. Inst. Fourier, 5, 1953-54, p.131-295)

[2] Existence et unicité des representations intégrales (Sém. Bourbaki, Dec. 1956).

[3] Potentiels sur un ensemble de capacité nulle-suites de potentiels (C.R. Ac. Sc. 244, 1957, p.1707).

[4] Sur les fondements de la théorie fine du potentiel (Sem. th. du potentiel 1, 1957).

[5] Forme abstraite du théorème de capacitabilité (Ann. Inst. Fourier 9, 1959, p.83 - 89).

[6] Sur les points d'effilement d'un ensemble. Application à l'étude de la capacité (Ann. Inst. Fourier, 9, 1959, 91-102).

[7] Démonstration non probabiliste d'un théorème de Getoor (Ann. Inst. Fourier 15/2, 1965, p. 409 - 413).

G. Choquet and P. A. Meyer, Existence et unicité des représentations intégrales dans les convexes compacts quelconques (Ann. Inst. Fourier, 14/2, 1964, p. 485 - 492).

C. Constantinescu, [1] Die heutige Lage der Theorie der harmonischen Räume.

[2] Some properties of the balayage of measures on a harmonic space (Ann. Inst. Fourier 17/1, 1967, p.273-293).

C. Constantinescu and A. Cornea [1] Ideale Ränder Riemannscher Flächen (Springer 1963).

[2] On the axiomatic of harmonic functions I and II (Ann. Inst. Fourier 13/2, 1963, p. 373-389-394).

[3] Compactifications of harmonic spaces (Nagoya Math. J. 25, 1965, p. 1 - 57).

J. Deny, [1] Un théorème sur les ensembles effilés (Annales Univ. de Grenoble 23 Math. Phys., 1947-48, p. 139 - 142).

[1bis] Le principe des singularités positives de G. Bouligand et la représentation des fonctions harmoniques positives dans un domaine (Revue Scientifique 85e année, fascicule 14, No.3279, 15 Août 1947, p.866-872).

[2] Sur les infinis d'un potentiel (C.R. Ac. Sc. 224, 1947,p.524).

[3] Les potentiels d'énergie finie (Acta Math. 82, 1950,p.107-183).

[4] Méthodes hilbertiennes en théorie du potentiel (Summer course, C.I.M.E. Stresa, 1969).

J. Deny and J.L. Lions, Les espaces du type de Beppo-Levi (Ann. I.F. 5, 1953-54, p.305-370).

J.L. Doob, [1] Semi-martingales and subharmonic functions (T.A.M.S. 77,1954, p. 86 - 121).

J.L. Doob, [2] Probability Methods applied to the first boundary value problem (Proc. third Berkeley Symp. 2, 1954-55, p.49 - 80).

[3] Conditional Brownian Motion and boundary limits of harmonic functions (Bull. Soc. Math. France, 85, 1957, p. 431 - 458).

[4] A non-probabilistic proof of the relative Fatou-theorem (Ann. Inst. Fourier, 9, 1959, p.293 - 300).

[5] Conformally invariant Cluster value theory (Illinois J. of Maths. 5, 1961, p.521 - 547).

[6] Boundary properties of functions with finite Dirichlet integrals (Ann. Inst. Fourier, 12, 1962, p. 573 - 621).

[7] Some classical function theory theorems and their modern versions (Colloque Paris-Orsay 1964, Ann. Inst. Fourier 15/1, 1965, p.113 - 136).

[8] Application to Analysis of a topological definition of smallness of a set (Bull. A.M.S. 72, 1966, p.579-600).

[9] Remarks on the boundary limits of harmonic functions (J. SIAM Numer. Anal. 3, n° 2, 1966, p.229 - 235).

E.B. Dynkin, [1] Markov Processes (English translation, Springer 1965, 2 volumes)

[2] Martin boundary and positive solutions of some boundary value problems (Colloque Paris-Orsay 1964, Ann. Inst. Fourier 15/1, 1965, p.275 - 282).

B. Fuglede, [1] Le théorème du minimax et la théorie fine du potentiel (Colloque Paris-Orsay, 1964 and Ann. Inst. Fourier 15/1, 1965, p. 65 - 87).

[2] Esquisse d'une théorie axiomatique de l'effilement et de la capacité (C.R. Ac. Sc. 261, 1965, p.3272).

[3] Quasi-topology and fine topology (Sém. potentiel Paris, mai 1966).

[4] The quasi-topology associated with a countably subadditive set function (to appear in Annales Inst. Fourier 21, 1971).

[5] Propriétés de connexion en topologie fine (to appear).

R.K. Getoor, Additive functionals of a Markov process (Lectures at Hamburg Univ. Summer 1964).

K.N. Gowrisankaran, [1] Extreme harmonic functions and boundary value problems (Ann. Inst. Fourier 13/2, 1963, p.307 - 356).

[2] Extreme harmonic functions and boundary value problems II (Math. Zeits. 94, 1966, p.256 - 270).

[3] Multiply harmonic functions (Nagoya Math. J. 28, Oct. 1966, p. 27 - 48).

K.N. Gowrisankaran, [4] Fatou-Naïm-Doob limit theorems in the axiomatic system of Brelot (Ann. Inst. Fourier, 16/2, 1966, p.455-467).

[5] On minimal positive harmonic functions (Sem. Th. potentiel 11, 1966-67).

Mrs. R. M. Hervé, [1] Recherches axiomatiques sur la théorie des fonctions surharmoniques et du potentiel (Ann. Inst. Fourier 12, 1962, p. 415 - 571).

[2] Un principe du maximum pour les sous-solutions locales d'une équation uniformément elliptique de la forme $Lu = -\sum_i \frac{\partial}{\partial x_i}\left(\sum_j a_{ij}\frac{\partial u}{\partial x_j}\right) = 0$ (Ann. Inst. Fourier 14,1964 p.493-508).

[3] Quelques propriétés des fonctions surharmoniques associès à une équation uniformément elliptique de la forme $Lu = -\sum_i \frac{\partial}{\partial x_i}\left(\sum_j a_{ij}\frac{\partial u}{\partial x_j}\right) = 0$ (Ann. Inst. 15/2,1965,p.215-223).

D. Hinrichsen, Randintegrale und Nukleare Funktionenraüme (Ann. Inst. Fourier 17/1, 1967, p.225 - 271).

G.A. Hunt, Markov Processes and Potentials, I,II,III (Ill. J. of Math (1) 1957, p.44 - 93, 316-369 (2) 1958, p. 151 - 213).

H.L. Jackson, Some results on thin sets in a half-plane (Ann. Inst. Fourier 20/2, 1970 to appear).

A. de la Pradelle, [1] Approximation et Caractère de quasi-analyticité dans la théorie axiomatique des fonctions harmoniques (Ann. Inst. Fourier, 17/1, 1967, p.383 - 399).

[2] Remarque sur la valeur d'un potentiel à support ponctuel polaire en son pôle en théorie axiomatique (Ann. Inst. Fourier, 19/1, 1969, p.275).

Ch-J. De La Vallée, Poussin, [1] Les nouvelles méthodes de la théorie du potentiel et le problème généralisé de Dirichlet (Act. Sc. et ind 578, Paris, Hermann 1937).

[2] Potentiel et problème généralisé de Dirichlet (Math. Gazette 22, 1938, p. 17 - 36).

[3] Le potentiel logarithmique; balayage et représentation conforme (Paris, Gauthier-Villars et Louvain Lib. Univ. 1949).

Mrs. J. Lelong-Ferrand, Etude au voisinage de la frontière des fonctions surharmoniques positives dans un demi-espace (Ann. Ecole. N. Sup 66, 1949, p. 125 - 159).

Mrs. L. Lumer-Naïm, [1] Sur le théorème de Fatou généralisé (Ann. Inst. Fourier 12, 1962, p. 623 - 626).

[2] Sur une extension du principe de Dirichlet en espace de Green (C.R. Ac. Sc. 255, 1962, p.1058.

Mrs. L. Lumer-Naïm, [3] $\mathcal{H}^p$ spaces of harmonic functions (Ann. Inst. Fourier 17/2, 1967, p.425 - 469).

R.S. Martin, Minimal positive harmonic functions (T.A.M.S. 49, 1941, p. 137 - 172).

P.A. Meyer, [1] Probabilités et Potentiels (Paris Hermann 1966, Act. Sc. et ind n° 1318, in English, Boston, Blaisdell publ. Company 1966).

[2] Processus de Markov (Lecture Notes no 26 and 77, Springer, 1967).

G. Mokobodzki, [1] Représentation intégrale des fonctions surharmoniques au moyen des réduites (Colloque Paris-Orsay 1964, and Ann. Inst. Fourier 15/1, 1965, p. 103 - 112).

[2] Cônes de potentiels et noyaux subordonnés (Summer course C.I.M.E., Stresa 1969).

A.D. Myskis, On the concept of boundary (Mat.Sbornik NS 25(67), 387-414 (1949) Translation in the collection of translations of the A.M.S. n° 51 (1951)).

L. Naïm, Sur le rôle de la frontière de R.S. Martin dans la théorie du potentiel (Ann. Inst. Fourier 7, 1957, p. 183 - 281).

M. Parreau, Sur les moyennes des fonctions harmoniques et analytiques et la classification des surfaces de Riemann (Ann. Inst. Fourier 3, 1951, p. 103 - 197).

F. Riesz, Sur les fonctions sousharmoniques et leur rapport à la théorie du potentiel (Acta. Math. Vol. 48, 1926, p.329-343 and Vol. 54, 1930, p. 321 - 360).

D. Sibony, [1] Allure à la frontière minimale d'une classe de transformations-théorème de Doob généralisé (Ann. Inst. Fourier 18/2, 1968, p. 91 - 120).

[2] Théorème de limites fines et problème de Dirichlet (Ann. Inst. Fourier 18/2, 1968, p. 121 - 134).

E. Smyrnelis, Allure des fonctions harmoniques au voisinage d'un point-frontière irrégulier (C.R. Ac. Sc. 267, July 1968, p.157).

G. Stampacchia, Le problème de Dirichlet pour les équations elliptiques du second ordre à coefficients discontinus (Colloque Paris-Orsay 1964, and Ann. Inst. Fourier 15/1, 1965, p.189-257).

E. Stein, On the theory of harmonic functions of several variables II Behaviour near the boundary (Acta. Math. 106, 1961, 137-174).

N. Toda, [1] Etude des fonctions méromorphes au voisinage d'un point-frontière irregulier (Bull. Sc. Math. 89, 1965, p. 93 - 102).

[2] Sur l'allure des fonctions méromorphes (Nagoya Math. J. 26 (1966), p. 173 - 181).

E. Tolsted, Limiting values of subharmonic functions (Bull. Am. Math. Soc. 1949, p. 636 - 647).

J. B. Walsh, Probability and a Dirichlet problem for multiply superharmonic functions (Ann. Inst. Fourier 18/2, 1968,p.221-279).

N. Wiener, [1] Certain Notions in Potential theory (J. of Math and Physics III, n° 1, Jan. 1924 or Publ. of the M.I.T, series II, n° 70, 1924).

[2] The Dirichlet problem (J. of Math. and Physics III, n° 3, April 1924 or Publ. of the M.I.T. series 11, n° 78, 1924).

[3] Note on a paper of O. Perron (J. of Math. and Phys. IV, n° 1, Jan. 1925 or Publ. of the M.I.T. series 11, n° 85,1925).

Index of Terminology